Optoelectronics for
Environmental Science

ETTORE MAJORANA
INTERNATIONAL SCIENCE SERIES
Series Editor:
Antonino Zichichi
European Physical Society
Geneva, Switzerland

(PHYSICAL SCIENCES)

Recent volumes in the series:

A Continuation Order Plan is available for this series. A continuation order will bring delivery
of each new volume immediately upon publication. Volumes are billed only upon actual ship-
ment. For further information please contact the publisher.

Optoelectronics for Environmental Science

Edited by

S. Martellucci

The Second University of Rome
Rome, Italy

and

A. N. Chester

Hughes Research Laboratories
Malibu, California

Plenum Press • New York and London

Library of Congress Cataloging in Publication Data

Course of the International School of Quantum Electronics on Optoelectronics for En-
vironmental Science (14th: 1989: Erice, Italy)
  Optoelectronics for environmental science / edited by S. Martellucci and A. N.
Chester.
    p.   cm.—(Ettore Majorana international science series. Physical sciences; v.
54)
  "Proceedings of the 14th course of the International School of Quantum Electronics
on Optoelectronics for Environmental Science, held September 3–12, 1989, in Erice,
Italy"—T.p. verso.
  Includes bibliographical references and index.
  ISBN-13: 978-1-4684-5897-8   e-ISBN-13: 978-1-4684-5895-4
  DOI: 10.1007/978-1-4684-5895-4
  1. Pollution—Measurement—Congresses. 2. Optical radar—Congresses. 3. Laser
spectroscopy—Congresses. 4. Optoelectronics—Congresses. I. Martellucci, S. II.
Chester, A. N. III. Title. IV. Series.
TD193.C68 1989   90-22278
628.5—dc20   CIP

Proceedings of the 14th course of the International School of
Quantum Electronics on Optoelectronics for Environmental Science,
held September 3–12, 1989,
in Erice, Italy

ISBN-13: 978-1-4684-5897-8

© 1991 Plenum Press, New York
Softcover reprint of the hardcover 1st edition 1991
A Division of Plenum Publishing Corporation
233 Spring Street, New York, N.Y. 10013

PREFACE

As we enter the nineties, there is worldwide awareness that the future of all mankind is inexorably linked by the world we share, and its response to man's activities.

Lasers and the optical sciences have brought powerful tools to measure and understand our environment. LIDAR (laser radar) and laser fluorescence allow us to measure atmospheric and oceanic pollutants, as well as industrial emissions, from many kilometers distance. And a variety of sensitive laser-based spectroscopic techniques permit the accurate analysis of heavy metals and other trace elements in the environment.

In September 1989, an international group of scientists met in Erice, Sicily, for the 14th Course of the International School of Quantum Electronics. This Course was devoted to "Optoelectronics for Environmental Science", and was ably directed by Prof. V. S. Letokhov of the USSR Institute of Spectroscopy and Prof. A. M. Scheggi of the C.N.R. Electromagnetic Waves Institute, Florence, Italy. This book gives the proceedings of that conference, which covered not only basic tutorial papers but also reports on the latest research results.

The first half of this volume describes the techniques used for direct "In-Situ Measurements" of the environment.

In "Techniques and Programs", four chapters and one extended abstract give tutorial discussions of the most important remote sensing techniques: LIDAR, laser fluorescence, and optical fiber sensors, plus a description of the Italian program in this area.

"Atmospheric Ozone" features three chapters, starting with the fundamentals of atmospheric ozone chemistry, and extending to a discussion of current data on the polar ozone hole. Other sections cover remote measurement of air velocity, the monitoring of industrial pollutants and process gases ("Other Atmospheric Sensing"), and the remote measurement of oil pollution and phytoplankton in the oceans ("Oceanographic Measurements").

The second half of the book covers "Laboratory Analytical Techniques", which complement field measurements. In "Heavy Metal Analysis", four chapters cover the trace analysis of heavy metals, followed by two chapters on the measurement of trace gases, including interesting long-term data from the analysis of polar ice ("Other Trace Elements"). In the last section "Techniques and Instrumentation" the final five chapters cover basic analytical techniques and their supporting instrumentation.

This International School was held under the auspices of the "Ettore Majorana" Centre for Scientific Culture, Erice, Italy. We acknowledge with gratitude the sponsorship of the Italian C.N.R., the Italian Ministry of Education, the Italian Ministry of Scientific and Technological Research, of the A.E.I. "Optoelectronics" Technical Group, the European Physical Society and the Sicilian Regional Government.

We are also grateful to Profs. Vladilen S. Letokhov and Anna Maria Scheggi for their able organization and direction of the Course, and to Mrs. Vanna Cammelli for her very specialized assistance in the successful organization of the School. Before concluding, we acknowledge Miss Roberta Colussi, who volunteered to retype and revise the entire volume, as well as the editor at Plenum Press London, Ms. Janie Curtis, for outstanding professional support.

The Directors of the International School of Quantum Electronics:

A.N. Chester   S. Martellucci
Vice-President and Director   Professor of Physics
Hughes Research Laboratories   The Second University of Rome
Malibu, California, U.S.A.   Rome, Italy

July 15, 1990

CONTENTS

OTHER TRACE ELEMENTS

TECHNIQUES AND INSTRUMENTATION

TECHNIQUES AND PROGRAMS

ATMOSPHERIC POLLUTION MONITORING USING LASER LIDARS

S. Svanberg

Department of Physics
Lund Institute of Technology
Lund, Sweden

I. INTRODUCTION

Advanced techniques are needed to monitor our threatened environment, to evaluate pollution levels and developmental trends. While tropospheric pollution has obvious manifestations in terms of health problems, water and soil acidification, and forest damage, human-induced stratospheric changes in the ozone layer, as evidenced by the occurrence of "ozone holes" at the polar caps, may have much more far-reaching consequences[1-6]. Laser spectroscopy provides powerful means for remote sensing of molecules in the atmosphere, yielding information on pollution levels as well as meteorological conditions. There are two major kinds of laser methods applicable in remote sensing[7-15]:

1) Long path absorption monitoring; and,

2) Lidar (Light detection and ranging), with subdivisions:
   - Fluorescence lidar
   - Raman scattering lidar
   - Mie scattering lidar
   - Differential absorption lidar (DIAL).

Long path absorption techniques are based on the same principles as spectrophotometry. However, by using laser beams of low divergence it is possible to use pathlength of several km instead of the 1 cm cuvette used in the chemical laboratory. A single-ended arrangement can be achieved by utilizing a corner cube retroreflector at the end of the light path and collecting the back-reflected light with a telescope. Tunable diode lasers and CW line-tunable CO_2 lasers are useful in this approach. Since all detected photons have travelled the same path no range resolution is obtained and only average concentrations can be determined. The technique has a powerful non-laser counterpart in DOAS (differential optical absorption spectroscopy)[16-18], where a distant high-pressure xenon lamp is used in combination with fast spectral scanning detection to overcome limitations posed by atmospheric scintillation.

Optoelectronics for Environmental Science, Edited by S. Martellucci and
A.N. Chester, Plenum Press, New York, 1990

Dial Equation:

$$P(\lambda_{On}.R)/P(\lambda_{Off}.R) = Exp(-2[\sigma(\lambda_{On}) - \sigma(\lambda_{Off})] \int_{R}^{R} N(R')dR')$$

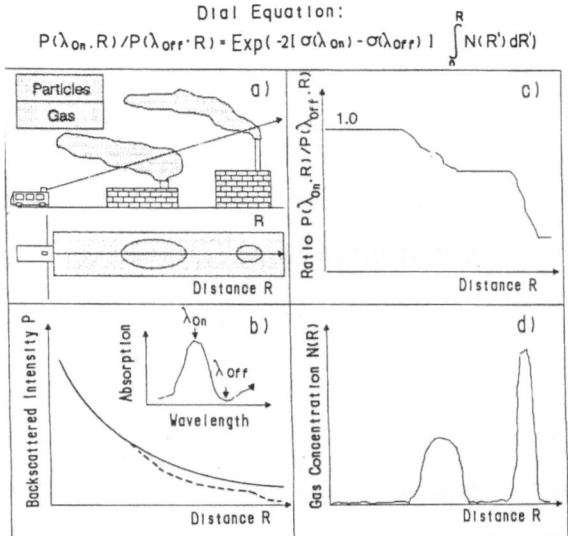

Fig. 1. Illustration of the principle of the
differential absorption lidar (DIAL)[23]:
(a) pollution measurement situation,
(b) back-scattered laser intensity for
the on- and off-resonance wavelengths,
(c) ratio (DIAL) curve, (d) evaluated
gas concentration.

II. THE LIDAR METHOD

In the lidar approach, a laser pulse is transmitted into the atmosphere and backscattered radiation is detected as a function of time by an
optical receiver, in a radar-like fashion. In the case of fluorescence
lidar a laser tuned to the absorption line of an atmospheric species is
used and fluorescence light is detected in the subsequent decay. Fluorescence lidar is a powerful technique for measurements at stratospheric
heights where the pressure is low and the fluorescence is not quenched by
collisions. The technique has been used extensively to monitor layers of
various alkali and alkaline earth atoms (Li, Na, K, Ca and Ca+) at a
height of about 100 km[19-21]. Raman lidar has many attractive features but
suffers from severe restrictions due to the weakness of the Raman scattering process. Only high gas concentrations can be detected at short ranges
using high-power lasers and optical Raman shift. Water vapor profiles can
be obtained in vertical soundings up to several km in height, and pressure
profiles up to tens of km are measurable using Raman signals from atmospheric N_2.

Mie scattering from particles provides strong signals allowing
mapping of the relative distribution of particles over large areas.
Stratospheric dust from volcanic eruptions can also be studied[22]. However,
since Mie scattering theory involves many normally inaccessible particle
parameters quantitative results are difficult to obtain. Mie scattering

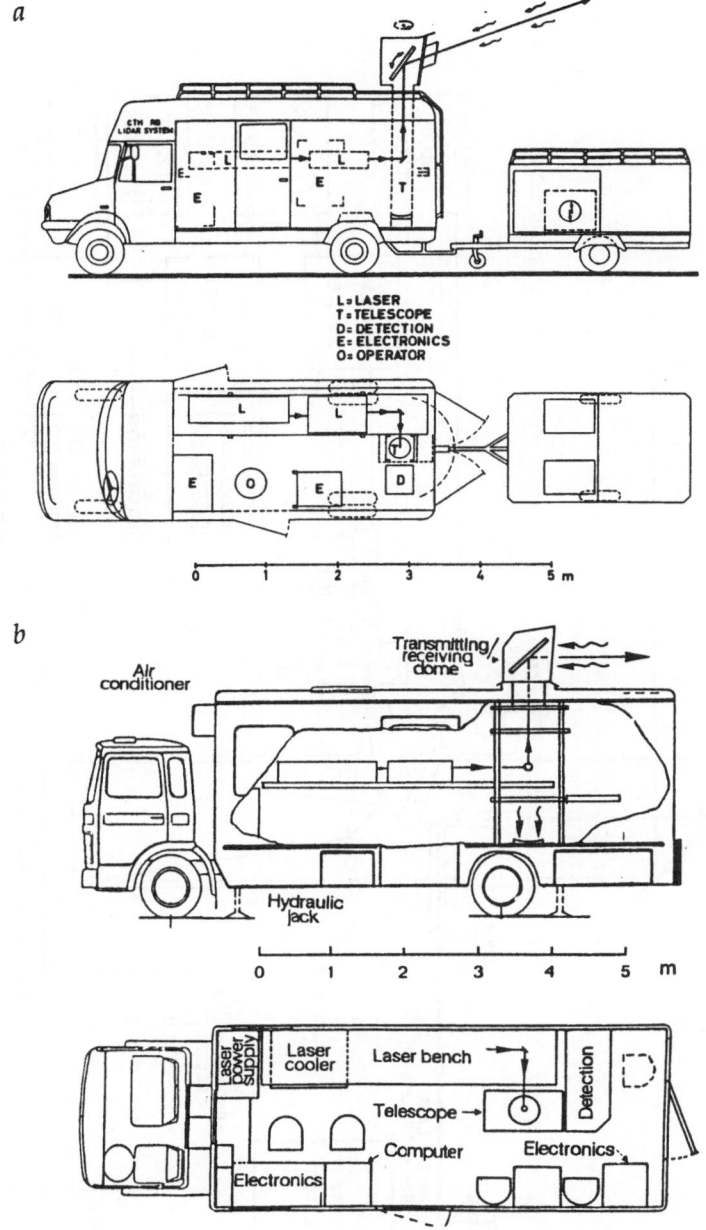

a

L = LASER
T = TELESCOPE
D = DETECTION
E = ELECTRONICS
O = OPERATOR

0 1 2 3 4 5 m

b

Air conditioner

Transmitting/receiving dome

Hydraulic jack

0 1 2 3 4 5 m

Laser power supply

Laser cooler

Laser bench

Telescope →

Detection

Computer

Electronics

Electronics

Fig. 2. Schematic views of two Swedish mobile lidar systems.
(a) System described in 1981 (See Ref. 24), (b) System
described in 1987 (see Ref. 23).

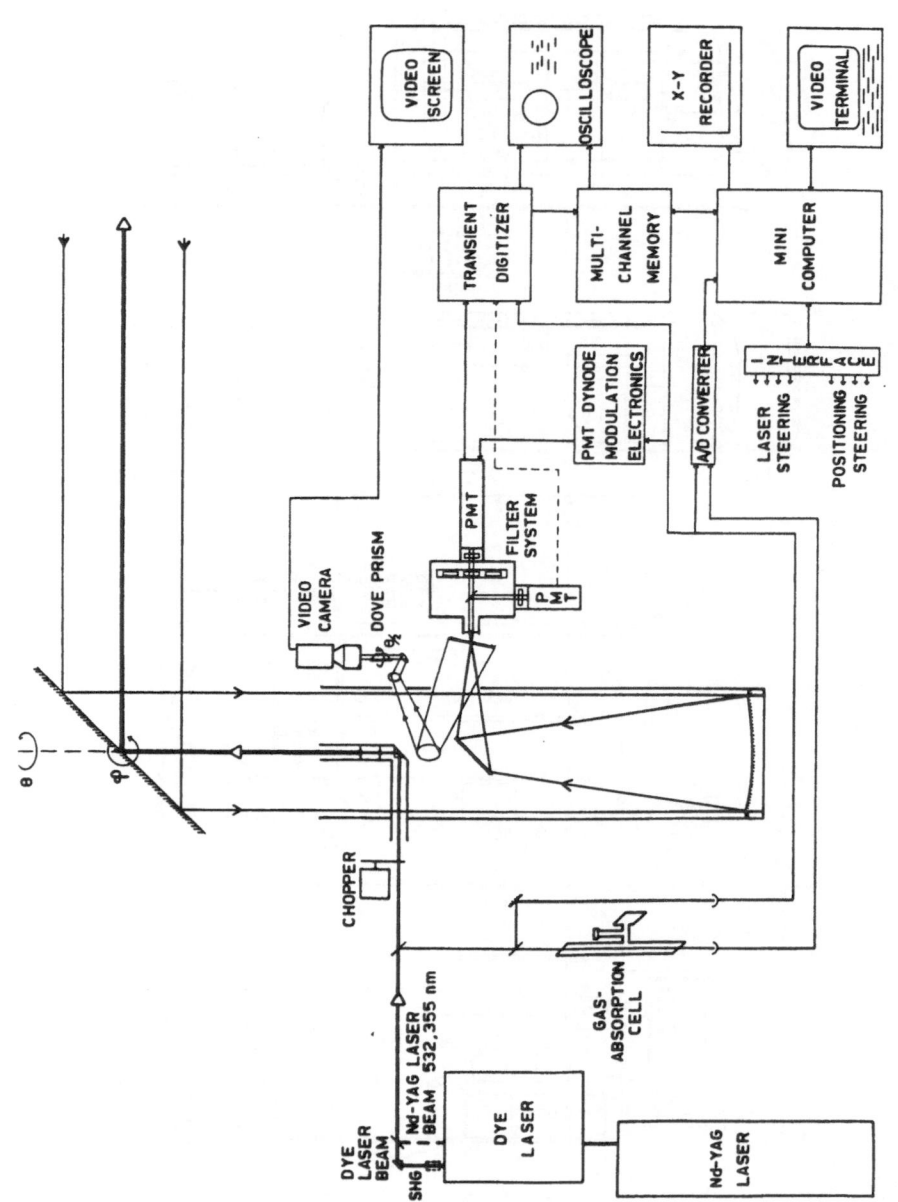

Fig. 3. Optical arrangement for a mobile lidar system[23].

6

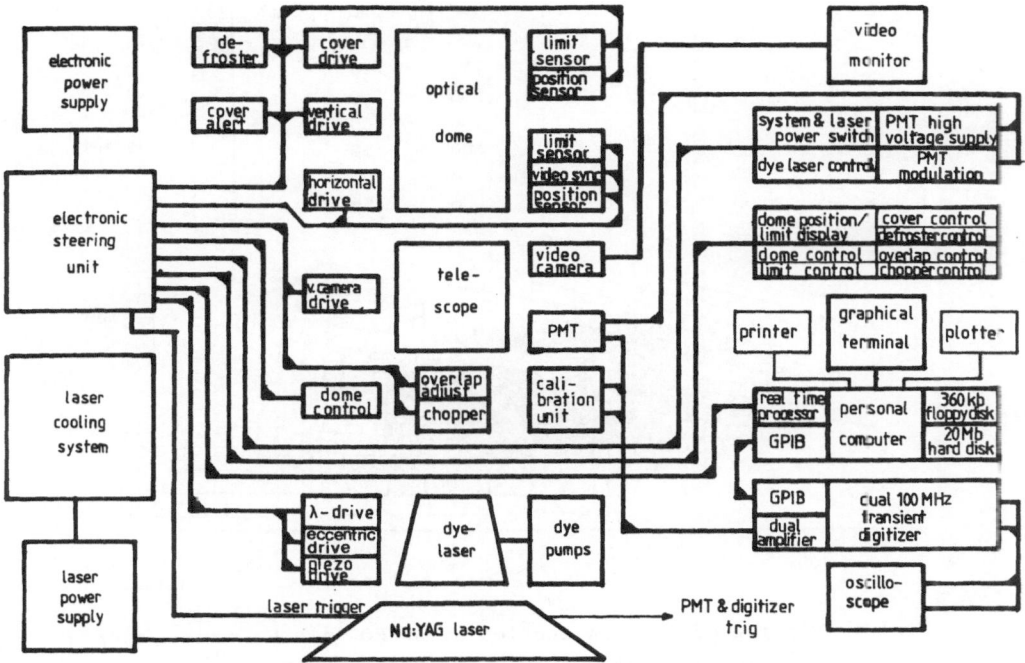

Fig. 4. Overview of the electronic system for a mobile lidar system[23].

is, however, extremely useful in providing the "distributed mirror" needed
in differential absorption lidar (DIAL).

III. DIFFERENTIAL ABSORPTION LIDAR

The principles for DIAL are schematically represented in Fig. 1 (Ref.
23). Laser light is alternately transmitted at a wavelength where the
species under investigation absorbs, and at a neighboring, off-resonant
wavelength. In the presence of an absorbing gas cloud the on-resonance
signal is attenuated through the cloud and the off-resonant one is not. By
dividing the two lidar signals by each other, most troublesome and unknown
parameters are eliminated and the gas concentration as a function of the
range along the beam can be evaluated. Mobile laser radar systems for
research and operational measurements have been constructed. In Figure 2
two Swedish systems are shown[23,24], one constructed at the Chalmers
Institute of Technology and the other one at the Lund Institute of
Technology. A schematic of the optical arrangements in the older system is
shown in Figure 3, while the electronic scheme of the newer system is
shown in Figure 4.

The DIAL technique is operational for important pollutants such as
SO_2, NO_2 and O_3, for which molecules the appropriate wavelengths can
readily be generated (see, e.g., Refs. 23-27). An example of data from a
vertical scan through a spreading SO_2 plume is given in Fig. 5. This type
of representation can be automatically generated using the system described
in Ref. 23. If the integrated concentration over the plume cross section

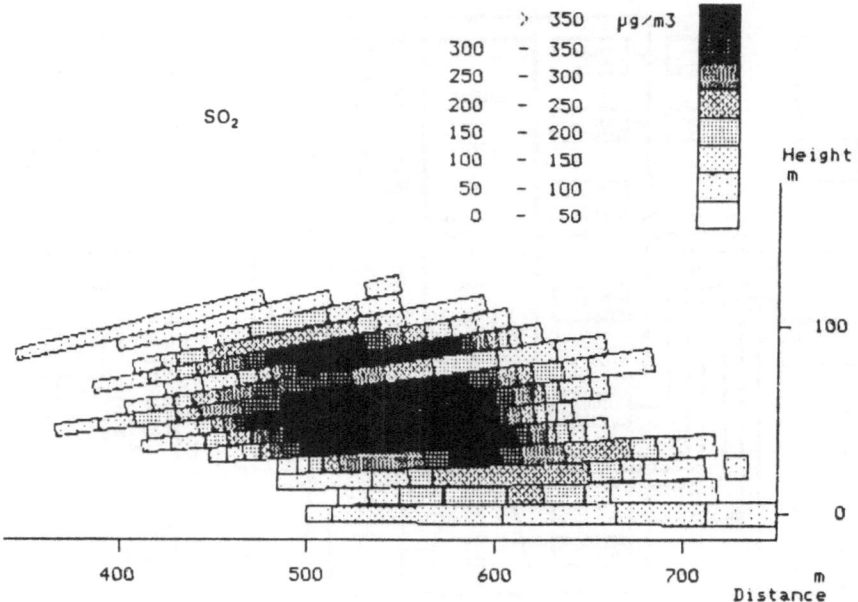

Fig. 5. Mapping of a cross section of an SO_2 plume from a paper mill obtained by a 20-minute dial measurement[23].

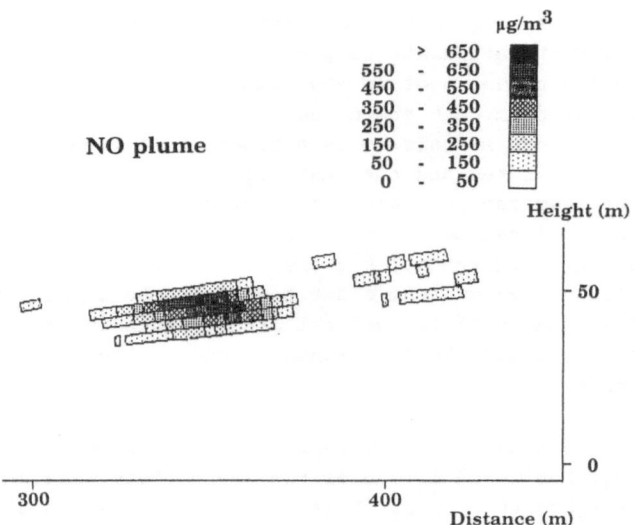

Fig. 6. NO plume mapping[28].

Fig. 7. Horizontal scan of Hg distribution over a
chlorine-alkali plant[29].

is multiplied by the wind velocity component vertical to the measurement
plane the total pollutant flux from the source is obtained. Pollutants
such as NO and Hg absorb at short UV wavelengths that more recently have
become accessible through nonlinear frequency conversion techniques in new
materials, such as β-barium-borate. Data from a vertical lidar scan for
NO, investigated at 226 nm are shown in Fig. 6 (Ref. 28).

Mercury is a troublesome pollutant generated from coal combustion,
chlorine-alkali and refuse incineration plants. Since it is present in the
atmosphere primarily in atomic form, it can be detected in very low concentrations (ppt) because its differential absorption is much higher than
that of molecules[29]. Mercury data from a vertical scan over a chlorine-
alkali plant are shown in Figure 7. Mercury is also an interesting geophysical tracer gas related to the occurrence of ore deposits, geothermal
reservoirs and seismic activity. An illustration of geophysical lidar
applications, centered mainly around mercury, is shown in Figure 8.

Great interest is presently being focused on the ozone molecule. A
steadily increasing concentration of tropospheric ozone is thought to be
related to the increasing damage to forests observed throughout Europe.
Ozone chemistry, dynamics and measurement techniques form an important
part of the inter-European EUROTRAC research program. Within this project,
a vertically sounding ozone lidar system is being developed in Lund[30]. The
system incorporates a 100 Hz repetition rate 500 mJ/pulse KrF excimer
system (λ=248 nm) which has an unstable resonator for high-quality beam
generation. A number of suitably located frequencies within the ozone
Hartley-Huggins absorption band can be generated by stimulated Raman
shifting in high-pressure H_2 cells[31,32]. Our system uses a horizontal

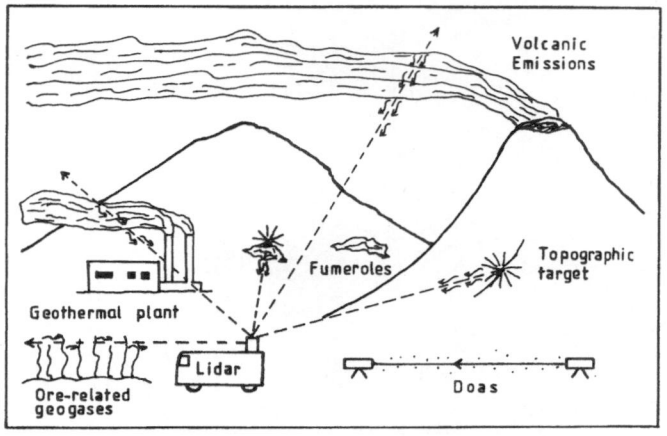

Fig. 8. Illustration of atmospheric remote sensing
related to geophysical research.

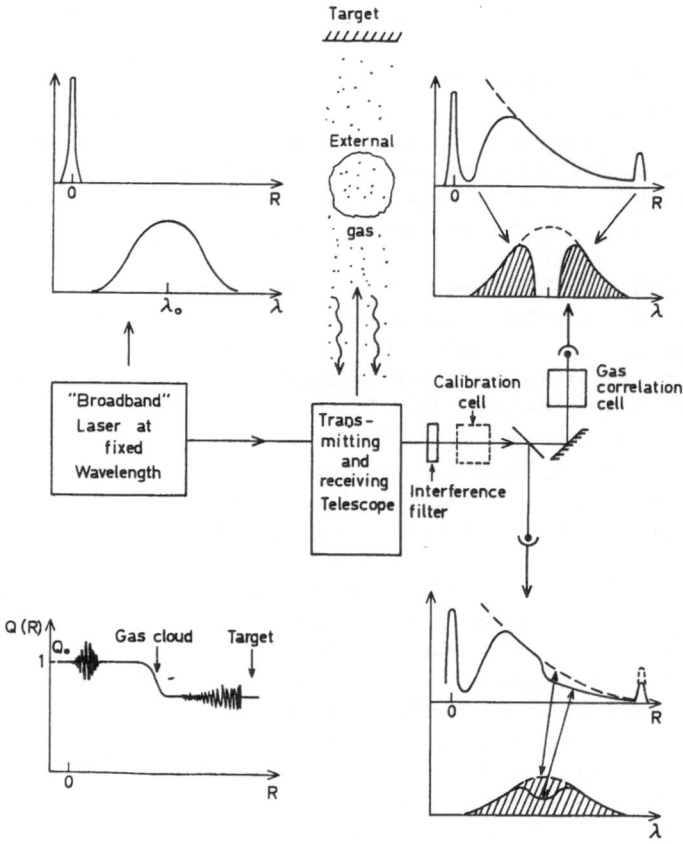

Fig. 9. The principle of gas correlation lidar [34].

30 cm diameter telescope placed inside the laboratory, and the transmitted and detected light beams are folded vertically by a large 45° mirror placed outside the building. A high-throughput spectrometer and multiple PMT tubes are used for simultaneous detection of two or more wavelengths. Ground-based DIAL systems for studies of the stratospheric ozone layer have been constructed by several groups in the search for anomalies and long-term trends[31,32]. Both NASA and ESA are planning space lidar systems providing global wind, temperature and pollution monitoring.

IV. DISCUSSION

Lidar systems are very powerful and provide unique three-dimensional information on atmospheric conditions. However, the systems also tend to be complex and expensive, which is a limiting factor in the widespread use of such systems for environmental management. Thus, there is a great need to develop simplified equipment. One such approach is gas correlation lidar[34] (see Figure 9). Here a rather crude laser system with a comparatively broad line-width is utilized. Since the laser wavelength is not sharp it covers both on-and off-resonance wavelengths at the same time. However, the information can be separated on the detection side by splitting the received radiation into two parts. One part is detected directly while the other part is first passed through a cell filled with an optically thick sample of the gas to be studied. In this way all the on-resonance radiation is filtered away leaving only the off-resonance radiation to be detected. In the direct channel the sum of the on- and off-resonance radiation is detected. Unknown factors are eliminated by dividing the signals. The simultaneous detection of the two signals also eliminates influences due to atmospheric turbulence and fluctuations due to changing reflectivity in airborne measurements using topographic targets. We are also investigating other aspects of gas correlation techniques, including multi-channel gas flow imaging[35,36], using ambient environmental optical radiation.

V. ACKNOWLEDGEMENTS

The author gratefully acknowledges fruitful cooperation with a large number of present and previous coworkers and graduate students in the field of atmospheric remote sensing. This work was supported by the Swedish Board for Space Activities, the National Swedish Environment Protection Board, the Swedish Natural Science Research Council, the Swedish Space Corporation and the Knut and Alice Wallenberg Foundation.

REFERENCES

1. W. Bach, J. Pankrath and W. Kellogg, eds., "Man's Impact on Climate", Elsevier, Amsterdam (1979)
2. R. Revelle, Sci. Amer., 247:35 (1982)
3. J. H. Seinfeld, "Atmospheric Chemistry and Physics of Air Pollution", Wiley, New York (1986)

4. R. P. Wayne, "Chemistry of Atmospheres", Clarendon Press, Oxford, (1985)

5. B. A. Trush, _Rep. Prog. Phys._, 51:1341 (1988)

6. R. S. Stolarski, Antarctic Ozone Hole, _Sci. Am._, 258/1:30 (1988)

7. D. A. Killinger and A. Mooradian, eds., "Optical and Laser Remote Sensing", Springer Series in Optical Sciences, vol. 39, Springer-Verlag, Heidelberg (1983)

8. R. M. Measures, "Laser Remote Sensing: Fundamentals and Applications", Wiley, New York (1984)

9. E. D. Hinkley, ed., "Laser Monitoring of the Atmosphere", Topics in Applied Physics, vol. 14, Springer-Verlag, Heidelberg (1976)

10. S. Svanberg, _Contemp. Phys._, 21:541 (1980)

11. S. Svanberg, Fundamentals of atmospheric spectroscopy, _in_: "Surveillance of Environmental Pollution and Resources by Electromagnetic Waves", T. Lund, ed., D. Reidel, Dordrecht (1978)

12. S. Svanberg, Laser technology in atmospheric pollution monitoring, in "Applied Physics - Laser and Plasma Technology", B. C. Tan, ed., World Science, p. 258, Singapore (1985)

13. W. B. Grant, Laser remote sensing techniques, _in_: "Laser Spectroscopy and its Applications", L. J. Radziemski, R. W. Solarz, and J. A. Paisner, eds., Marcel Dekker, p. 565, New York (1987)

14. T. Kobayashi, _Rem. Sens. Rev._, 3:1 (1987)

15. R. M. Measures, ed. "Laser Remote Chemical Analysis", Wiley-Interscience, New York (1988)

16. U. Platt, D. Perner and H. W. Patz, _J. Geophys. Res._, 84:6329 (1979)

17. U. Platt and D. Perner, Measurements of atmospheric trace gases by long path differential UV/visible absorption spectroscopy, in Ref. (7)

18. H. Edner, A. Sunesson, S. Svanberg, L. Uneus and S. Wallin, _Appl. Opt._, 25:403 (1986)

19. M. L. Chanin, Rayleigh and resonance sounding of the stratosphere and mesosphere, in Ref. (7)

20. C. Granier and G. Megie, _Planet. Space Sci._, 30:169 (1982)

21. K. H. Fricke and U. V. Zahn, _J. Atm. Terr. Phys._, 47:499 (1985)

22. M. P. Mc Cormick, T. J. Swisser, W. H. Fuller, W. H. Hunt, and M. T. Osborn, _Geofisica Internacional_, 23-2:187 (1984)

23. H. Edner, K. Fredriksson, A. Sunesson, S. Svanberg, L. Unéus and W. Wendt, _Appl. Opt._, 26:4330 (1987)

24. K. Fredriksson, B. Galle, K. Nystrom and S. Svanberg, _Appl. Opt._, 20:4181 (1981)

25. K. Fredriksson and S. Svanberg, Pollution monitoring using Nd:YAG based lidar systems, in Ref. (7)

26. K. Fredriksson and H. M. Hertz, _Appl. Opt._, 23:1403 (1984)

27. A. L. Egeback, K. Fredriksson and H. M. Hertz, _Appl. Opt._, 23:722 (1984)

28. H. Edner, A. Sunesson and S. Svanberg, _Opt Letters_, 12:704 (1988)

29. H. Edner, G. W. Faris, A. Sunesson and S. Svanberg, _Appl. Opt._, 28:921 (1989)

30. H. Edner, P. Ragnarsson, S. Svanberg and E. Wallinder, "Vertical ozone probing with Lidar", Nordic Symposium on Atmospheric Chemistry, Helsinki, Dec. 6-8 (1989)

31. O. Uchino, M. Togunaga, M. Maeda and Y. Miyazoe, <u>Opt. Lett.</u>, 8:347 (1983)
32. J. Werner, K. W. Rothe and H. Walther, <u>Appl. Phys.</u>, B32:113 (1983)
33. G. Megie, G. Ancellet and J. Pelon, <u>Appl. Opt.</u>, 24:3454 (1985)
34. H. Edner, S. Svanberg, L. Uneus and W. Wendt, <u>Opt. Lett.</u>, 9:493 (1984)
35. P. S. Andersson, S. Montàn and S. Svanberg, <u>IEEE J. Quant. Electron.</u>, QE-23:1798 (1987)
36. P. Ragnarsson, "Spectroscopic imaging of effluent gases", Diploma paper, Lund Reports on Atomic Physics LRAP-83 (1988)

LASER FLUORESCENCE SPECTROSCOPY IN ENVIRONMENTAL MONITORING

S. Svanberg

Department of Physics
Lund Institute of Technology
Lund, Sweden

I. INTRODUCTION

Fluorescence spectroscopy has long been used for analytical and diagnostic purposes[1-4]. Using UV laser sources, the techniques of laser-induced fluorescence (LIF) have become particularly powerful. The LIF process in large molecules, such as biological ones, is schematically illustrated in Figure 1 (Ref. 5). The ground as well as the excited electronic levels are broadened by vibrational motion and interactions with surrounding molecules. Thus, absorption occurs in a broad band allowing a fixed-frequency UV laser, such as a nitrogen laser (λ=337 nm), an excimer laser (XeCl, λ=308 nm; XeF, λ=351 nm) or a frequency-tripled Nd:YAG laser (λ= 355 nm) to be used for the excitation. A radiationless relaxation to the bottom of the excited band then occurs on a pico-second time scale. The molecules remain here for a typical lifetime of few nanoseconds. Fluorescent light is released in a red-shifted broad band, which is frequently rather structureless. Internal conversion and transfer of energy to surrounding molecules are strongly competing radiationless processes.

Laser-induced fluorescence has an interesting potential for remote sensing of environmental parameters. For quite some time, hydrospheric pollution monitoring has been performed with airborne laser-based fluorosensors. Different kinds of oil can be identified by their fluorescence properties. Other pollutants and algal bloom patches can also be studied[6-8]. Using the blue-green transmission window of water, bathymetric measurements of sea depths can also be performed[9,10]. Some laboratory and field work performed by our group is described in Refs. 11-18. The field of laser-based hydrospheric monitoring is covered in Refs. 19-21. Laser-induced fluorescence has also been used by American[22,23], Italian[24] and Swedish groups for studies of land vegetation. Some work performed by our group is reported in Refs. 11 and 25.

The purpose of the present paper is to give a brief presentation of our work within the field of environmental monitoring using laser fluores-

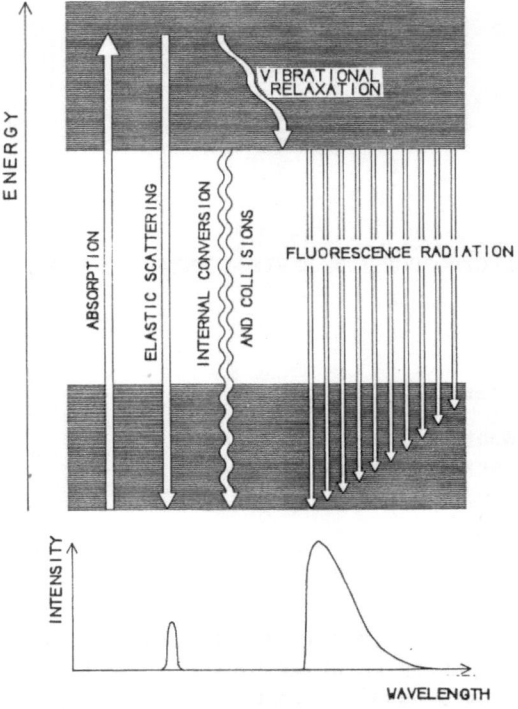

Fig. 1. Schematic diagram of the LIF process in large molecules[5].

cence spectroscopy. We will give examples from laboratory measurements and
field work. Finally, some examples of LIF studies using similar techniques
outside the environmental field will be mentioned.

II. LABORATORY MEASUREMENTS

The starting point for assessing the potential of LIF for environmental studies is laboratory studies of constituents of environmental
interest, e.g. mineral oils, phytoplankton and land vegetation. A suitable
experimental set-up for such studies is shown in Figure 2 (Ref. 11) . An
arbitrary sample is excited with a pulsed UV laser and the fluorescence is
dispersed in a spectrometer and detected. A scanning monochromator can be
used in conjunction with a photomultiplier tube and a boxcar integrator.
Better, an optical multichannel analyzer system with a gated and intensified

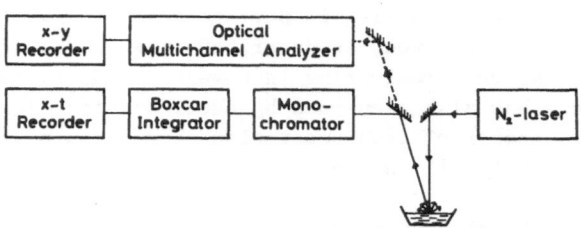

Fig. 2. Laboratory set-up for LIF studies[11].

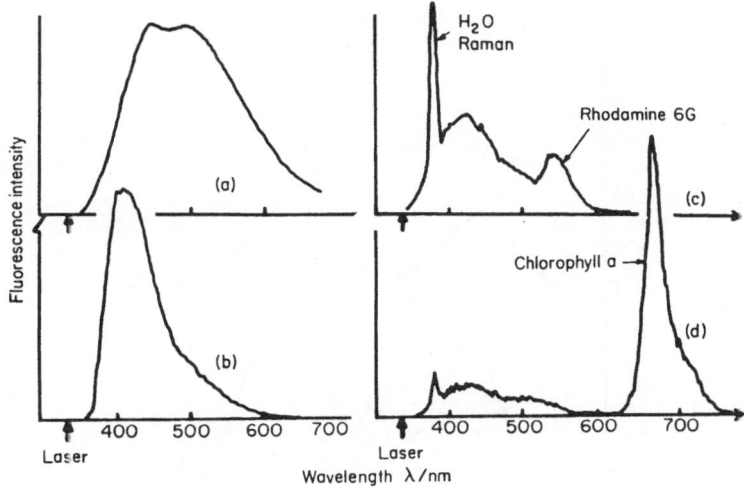

Fig. 3. Fluorescence spectra of various samples for nitrogen
laser excitation (λ=337 nm)[26]. (a) Crude oil, Abu Dhabi;
(b) river water down-stream from a sodium sulfite pulp
mill (the fluorescence is about 30 times stronger than
that from clean water); (c) seawater, Kattegat, containing 3 μg/l Rhodamine-6G dye, added as a tracer for
hydrological studies; (d) the green algae "Clorella
ovalis" Butcher in seawater.

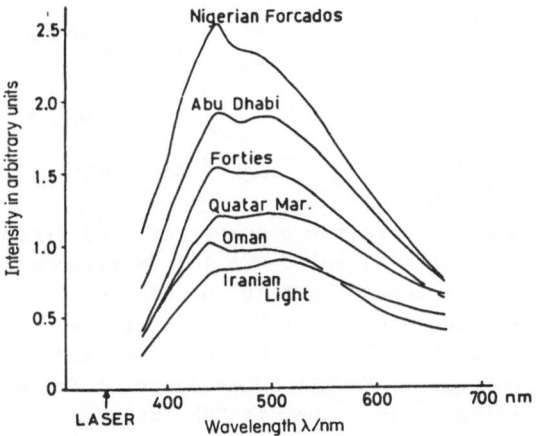

Fig. 4. LIF spectra for different crude oils[11].
The excitation wavelength is 337 nm.
The data are spectrally corrected.

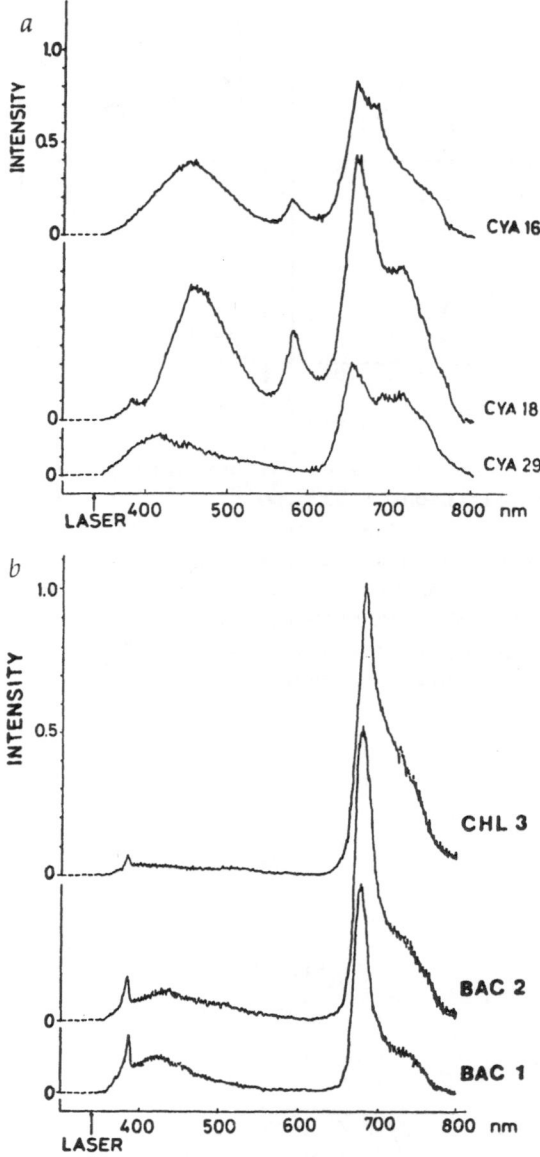

Fig. 5. Fluorescence spectra for a number of: (a) fresh water; and, (b) seawater algae[11]. CYA 16: "Merismopedia punctata Meyen (Cyanophyceae)", CYA 18 and CYA 29: Two taxonomically closely related species of "Oscillatoria agardhii" Gom. Var. ("Cyanophyceae"), CHL 3: "Chlorella ovalis" Butcher ("Chlorophyceae"), BAC 2: "Phaeodactylum Tricornutum" Bohlin ("Bacillariophyceae"), BAC 1: "Skeletonema costatum" (Grev.) Cl. ("Bacillariophyceae). The excitation wavelength is 337 nm and the data are spectrally corrected.

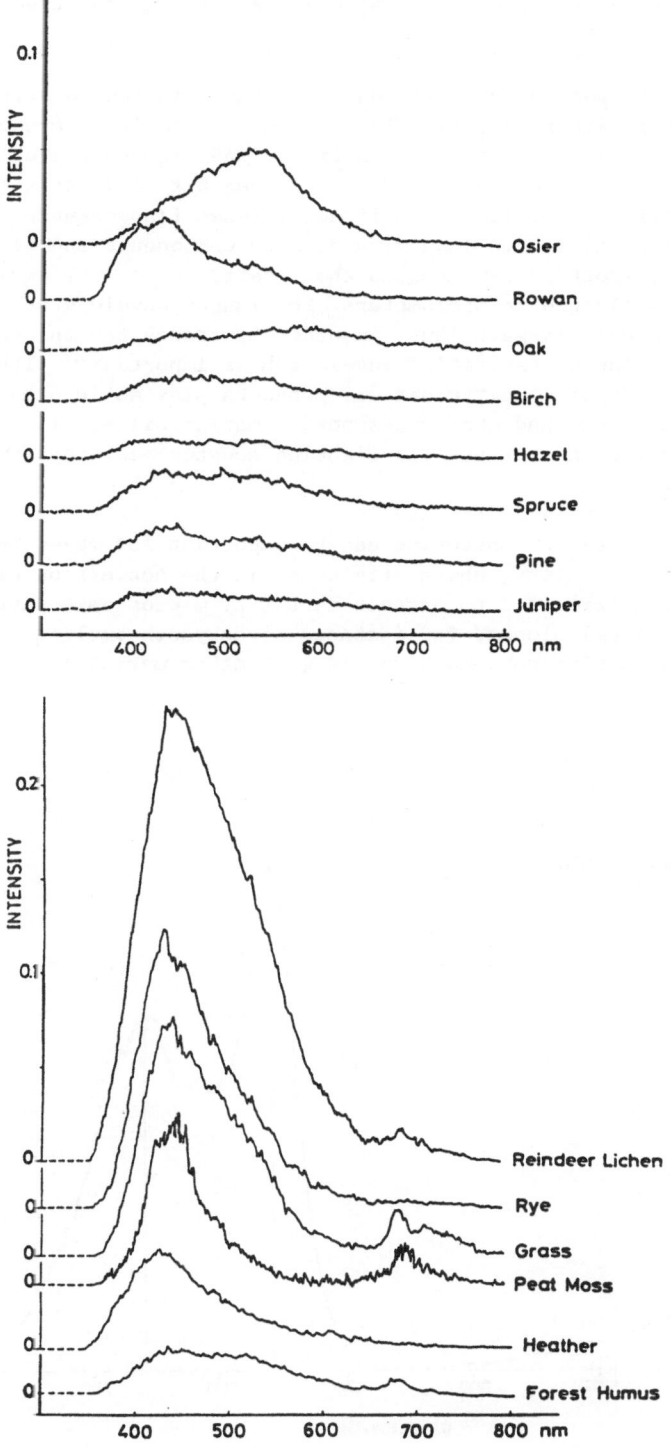

Fig. 6. Fluorescence spectra from different trees and plants[11]. Nitrogen laser excitation was employed. Data are spectrally uncorrected.

linear diode array is employed to capture the full LIF spectrum for each laser shot.

Examples of spectra for different substances in the aquatic environment are shown in Figure 3 (Ref. 26), including data for a crude oil, polluted river water, seawater with a tracer and a green algae. Spectra of different crude oils are found in Figure 4 (see Ref. 11). As a rule, light petroleum fractions exhibit blue-shifted, intense fluorescence while heavier fractions also have longer wavelength components and fluoresce more weakly. At short UV wavelengths the penetration of the exciting light into the oil is limited to micrometers. For longer wavelengths the penetration depth is larger. Thus, in order to assess the thickness of an oil film the choice of excitation wavelength is important[27]. The fluorescence characteristics of different oil products play an important role in airborne measurements and the assessment of marine oil spills in the decision regarding the correct oil-fighting counter-measures to be implememented.

Algal fluorescence monitoring can be important for measuring the total marine productivity, which originates in the conversion of solar energy CO_2 and nutrients into organic matter by microscopic phytoplankton. Recently, huge algal blooms, for instance of chrysochromulina polylepis, leading to devastating consequences for most other marine life forms have

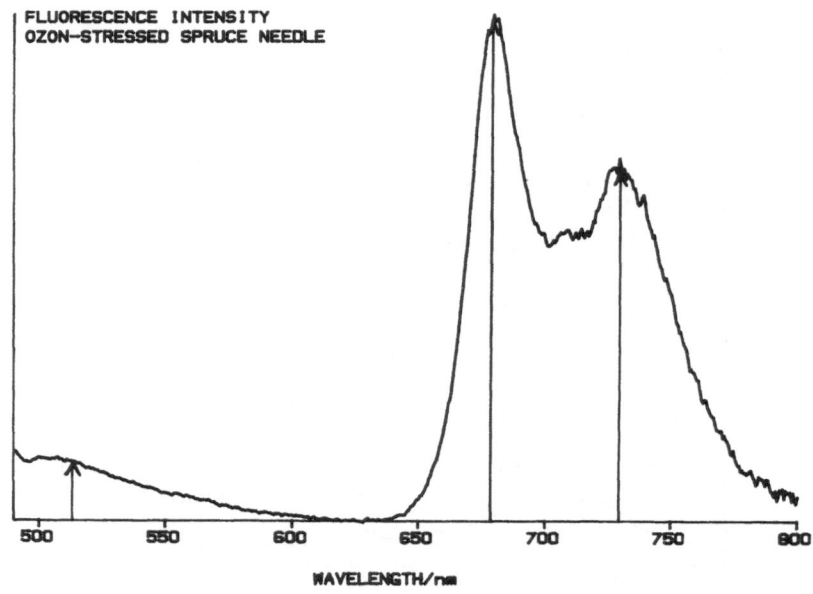

Fig. 7. LIF spectrum of spruce needles from a tree exposed to ozone[25]. The excitation wavelength was 405 nm, and the curve is spectrally corrected.

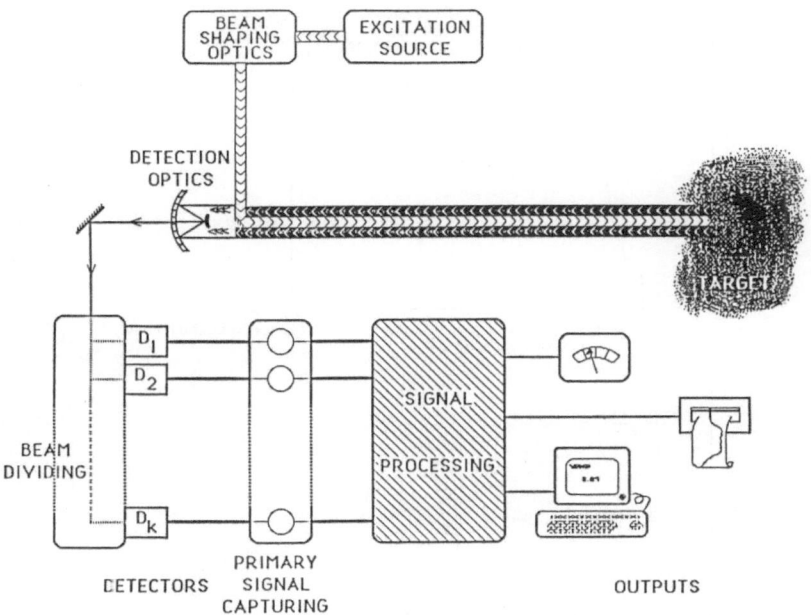

Fig. 8. General set-up for remote sample characterization based on LIF[17].

occured due to eutrophication of coastal waters. Some classes of algae exhibit LIF spectra with certain characteristic features in addition to the dominating peak at 685 nm due to chlorophyll a. Examples of algae spectra are given in Figure 5 (Ref. 11).

Land vegetation can be efficiently characterized by multispectral reflectance measurements using space-borne sensors in satellites such as LANDSAT, SPOT and ERS-1[28] . An "active" remote sensing technique such as LIF might, in certain circumstances, complement "passive" reflectance monitoring. We have studied the LIF properties of many species under 337 nm excitation, as illustrated in Figure 6 (Ref. 11). In many cases, the chlorophyll peak is absent since the short-wavelength light is absorbed in the leaf "wax" layer, which dominates the spectrum. For longer excitation wavelengths, penetration to the chlorophyll occurs, as shown in Figure 7 (Ref. 25) for spruce needles, excited at 405 nm, in an attempt to assess forest damage due to ozone exposure using LIF techniques.

III. FLUOROSENSOR CONSIDERATIONS

Once characteristic spectral features for materials of environmental interest have been established, the successful practical application of the LIF techniques, e.g. in airborne remote sensing, will depend on the signal-to-noise ratio obtainable in recording faint fluorescence phenomena in the presence of the environmental optical background radiation. The basic elements of a fluorosensor are shown in Figure 8 (Ref. 17). A critical

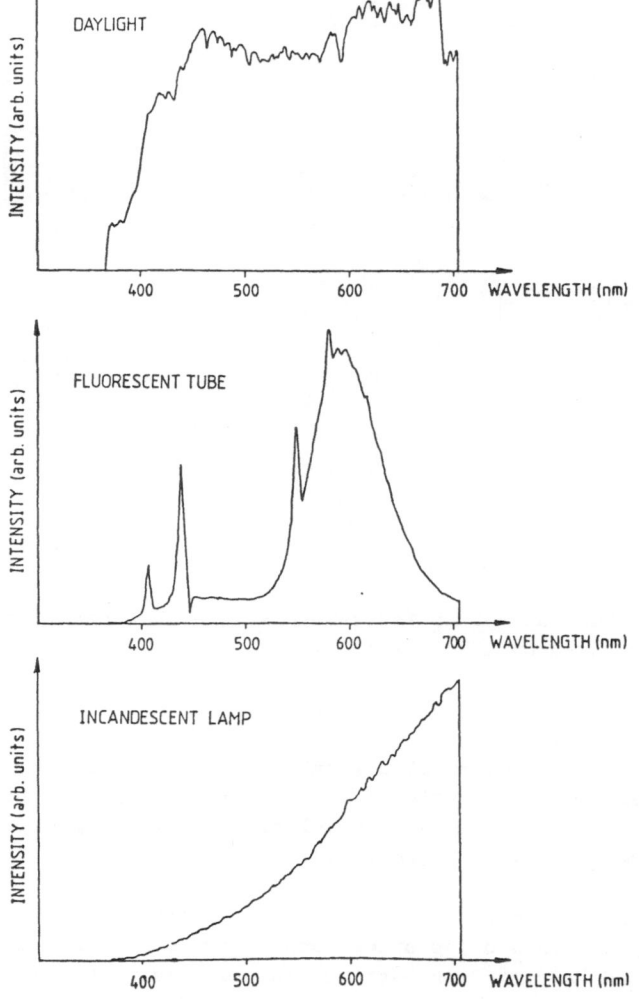

Fig. 9. Spectra of daylight, a fluorescent tube
and a tungsten lamp[17].

discussion of the different subsystems can be found in Ref. 17, considering transmitter, optics, electronics etc. In night-time operation the
system is basically limited only by the signal photon statistics, while
day-time operation is largely a question of background suppression, using
short grating times and narrow field-of-view optics. In Figure 9 (Ref. 17)
the ambient daylight spectral distribution is compared with those of fluorescent tube lamps and incandescent lamps, relevant in, e.g. industrial or
medical LIF applications. In certain cases a low-cost pulsed UV lamp could
be considered as a replacement for the laser. Lamps have a much longer
pulse duration (>10 μs) than lasers, but higher pulse powers. This
difference is illustrated in Figure 10 (Ref. 17), where oil fluorescence at
420 nm is detected at a distance of 70 meters under day-time and night-time

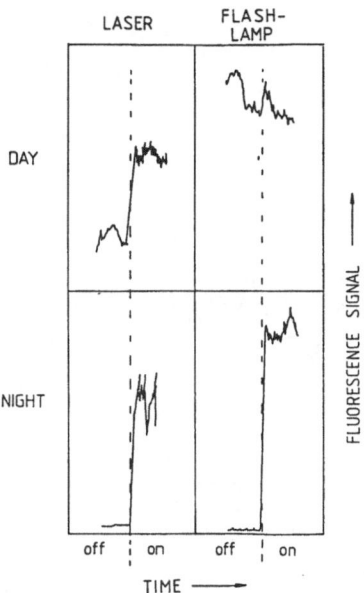

Fig. 10. Comparison between day- and night-
time LIF spectra from oil, recorded
at 70 meters distance from the
target. A pulsed laser or a UV
flashlamp was employed[17].

conditions. As can be seen, the lamp is superior during the night-time
while the laser is indispensable during the daytime.

Another important aspect of environmental fluorescence characterization is illuminated in Figure 11 (Ref. 13). The substance of interest must
exhibita sufficiently high contrast in relation to the surrounding, normal
material. Data from a marine field test are given in Figure 11 (Ref. 13).
A vertically sounding fluorosensor installed on a research vessel was
employed. In Figure 11a the LIF intensity at 465 nm from a discharge of
detergent solution into the sea water is shown. The specific signal is
superimposed on a strong background signal due to the integrated sea water
fluorescence. In the same way, the LIF intensity from a thin oil layer of
strong intrinsic intensity will not necessarily strongly dominate over the
intrinsically weak sea water fluorescence integrated over a layer thickness of the order of 1 m. The ratio signal I(565 nm)/I(465 nm) for an oil
discharge is shown in Fig. 11b, with a contrast to the surrounding water of
less than 2. The fluorescence integration of the natural water is difficult
to simulate in the laboratory. These phenomena are obviously very important
when the detection of sunken oil is considered. Experiments and conclusions
on this topic are presented in Ref. 15.

An example of a scenario for airborne oil-slick characterization is
given in Fig. 12 (Ref. 17). Here, a simple laser fluorosensor to complement
a side-looking airborne radar (SLAR) system is considered. The microwave-

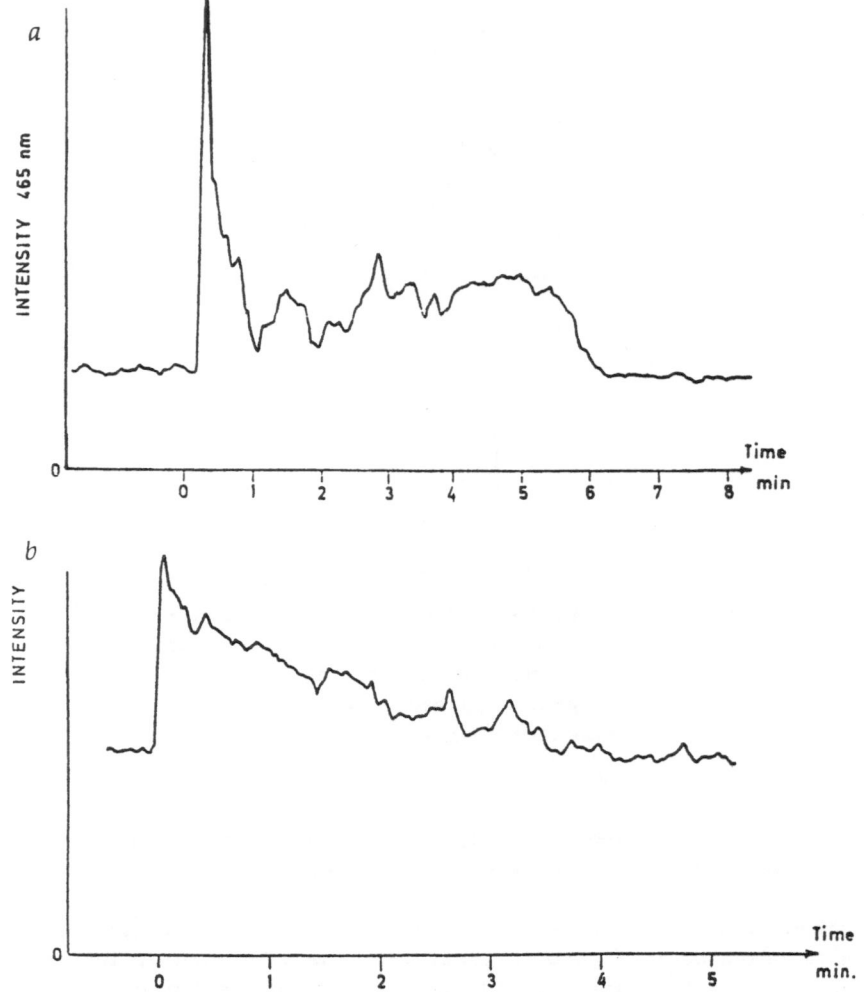

Fig. 11. Data from a marine field test using a ship-mounted nitrogen
laser fluorosensor for vertical sounding. (a) LIF signal at
465 nm connected to a discharge of detergent solution into
the water; (b) monitoring of oil floating on seawater. The
ratio I(565 nm)/I(465 nm) was measured. Since a ratio is
dimension-less, it is independent of geometrical factors
such as the presence of waves[13].

based system has a wide action range and all-weather capability in oil
slick detection based on a reduced sea-clutter back-scattering signal from
areas devoid of short-wavelength capillary water waves. Once a slick is
detected, the fluorosensor can characterize the oil type using a laser of
low pulse energy at short range. Construction considerations for such a
system are presented in Ref. 18.

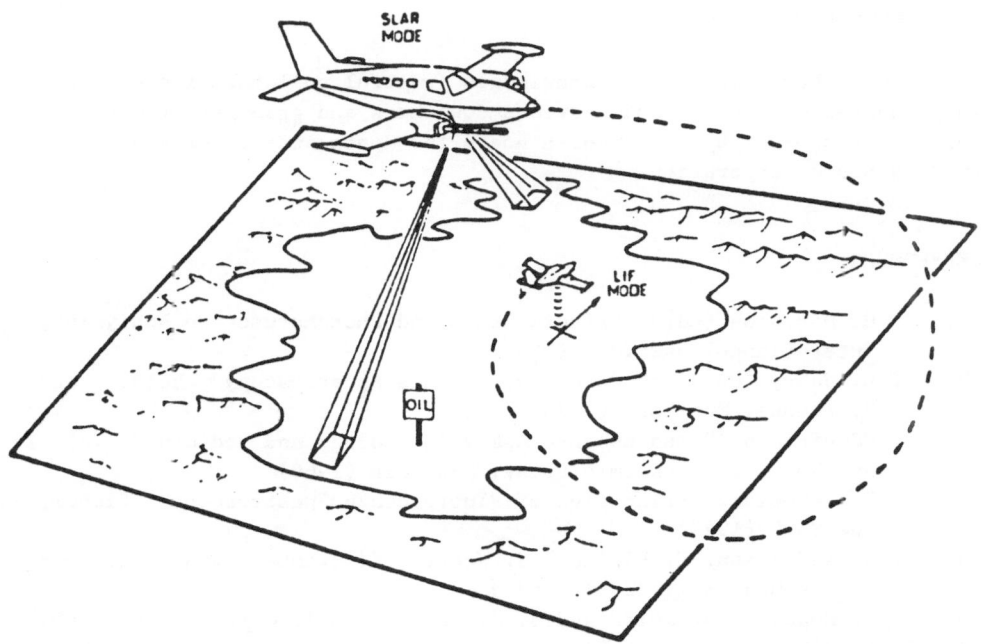

Fig. 12. Measurement scenario for an airborne marine surveillance system[17].

IV. DISCUSSION

Fluorescence techniques can complement passive multispectral reflectance remote sensing in environmental monitoring. Several functioning demonstration systems have been constructed. For a more widespread use of LIF techniques in these contexts the development of efficient, light-weight and cheap pulsed UV lasers is highly desirable. Similarly, cheaper image-intensified array detectors would favorably influence the development in this field.

Fluorescence monitoring techniques can obviously also be used for industrial purposes. Since oils possess very strong fluorescence properties the presence or absence of thin surface layers on, e.g., sheet metal can be determined. Thus, surface cleanliness, corrosion protective measures, etc., can be assessed with an industrial fluorosensor. Laboratory and field work on industrial laser-induced fluorescence are reported in Refs. 27, 29, 30.

Finally, laser-induced fluorescence is finding many interesting applications in medicine, and many of the techniques used for environmental and industrial monitoring can be adopted for cancer diagnosis, using tumor-seeking agents, and for localizing atherosclerotic plaques in human vessels. These aspects have been pursued actively within our laboratory during the last few years and reviews of this work can be found in Refs. 31-34.

V. ACKNOWLEDGEMENTS

The author gratefully acknowledges fruitful collaboration with a large number of present and previous coworkers and graduate students. This work was supported by the Swedish Board for Space Activities and the Swedish Space Corporation.

REFERENCES

1. D. H. Hercules (ed.), "Fluorescence and Phosphorescence Analysis", Interscience, New York (1966).

2. E. L. Wehry (ed.), "Modern Fluorescence Spectroscopy", Vols. 1 and 2, Plenum, New York (1976)

3. S. Udenfriend, "Fluorescence Assay in Biology and Medicine", Vol. I and Vol. II , Academic Press, New York (1969)

4. J. R. Lakowicz, "Principles of Fluorescence Spectroscopy", Plenum, New York (1983)

5. P. S. Andersson, E. Kjellen, S. Montán, K. Svanberg and S. Svanberg, _Lasers Med. Sci._, 2:41 (1986)

6. F. E. Hoge, R. N. Swift and J. K. Yungel, _Appl. Opt._, 25:48 (1986)

7. R. A. O'Neill, L. Buja-Bijunas and D. M. Rayner, _Appl. Opt._, 19:863 (1980)

8. G. A. Capelle, L. A. Franks and D. A. Jessup, _Appl. Opt._, 22:3382 (1983)

9. H. H. Kim, _Appl. Opt._, 16:46 (1977)

10. F. E. Hoge, R. N. Swift and E. B. Frederick, _Appl. Opt._, 19:871 (1980)

11. L. Celander. K. Fredriksson, B. Galle and S. Svanberg, Investigation of laser-induced fluorescence with applications to remote sensing of environmental parameters, Goteborg Institute of Physics Reports GIPR-149, CTH, Goteborg (1978)

12. K. Fredriksson, B. Galle, K. Nyström, S. Svanberg and B. Oström, Underwater laser-radar experiments for bathymetry and fish-school detection, Göteborg Institute of Physics Reports GIPR-162, CTH, Göteborg (1978)

13. K. Fredriksson, B. Galle, K. Nyström, S. Svanberg and B. Öström, Marine laser probing - results from a field test, Medd. Fr. Havsfiskelaboratoriet 245, Swedish Fishery Board, Lysekil (1979)

14. B. Galle, T. Olson an S. Svanberg, The Fluorescence properties of jelly-fish, Goteborg Institute of Physics Reports GIPR-181, CTH, Göteborg (1979) (in Swedish)

15. P. S. Andersson, S. Montán and S. Svanberg, Oil-slick characterization using an airborne fluorosensor - construction considerations, Lund Reports on Atomic Physics LRAP-45, LTH, Lund (1985)

16. P. S. Andersson, S. Montán and S. Svanberg, Flashlamps for remote fluorescence characterization of oil slicks, Lund Reports on Atomic Physics LRAP-57, LHT, Lund (1986)

17. P. S. Andersson, S. Montán and S. Svanberg, _Appl. Phys._, B44:19 (1987)

18. P. S. Andersson, S. Montán and S. Svanberg, Fluorosensor for remote characterization of marine oil-slicks, Intern. Coll. on Remote

Sensing of Pollution of the Sea, Oldenburg, March 31 - April 3 (1987)

19. R. M. Measures, "Laser Remote Sensing: Fundamentals and Applications", Wiley, New York (1984)

20. F. E. Hoge, Ocean and terrestrial lidar measurements, in: "Laser Remote Chemical Analysis", R. M. Measures, ed., Wiley-Interscience, New York (1988)

21. S. Svanberg, Contemp. Phys., 21:541 (1980)

22. F. E. Hoge, R. N. Swift and J. K. Yungel, Appl. Opt., 22:2991 (1983)

23. E. W. Chappelle, F. M. Wood, W. W. Newcomb and J. E. McMurtrey, Appl. Opt., 24:74 (1985)

24. F. Castagnoli, G. Cecchi, L. Pantani, I. Pippi, B. Radicati and P. Mazzinghi, A fluorescence lidar for land and sea remote sensing, SPIE, 663:212 (1986)

25. S. Andersson-Engels, K. Callander and B. Galle, Investigation of the possibilities to use laser-induced fluorescence to map conifer forest damage caused by ozone, IVL Report L88/146, IVL, Göteborg (1988)

26. S. Svanberg, Environmental diagnostics, in: "Trends in Physics", M. M. Woolfson, ed., Adam Hilger, Bristol (1979)

27. P. Herder, T. Olsson, E. Sjöblom and S. Svanberg, Monitoring of surface layers using laser-induced fluorescence, Lund Reports on Atomic Physics LRAP-9, LTH, Lund (1981)

28. H. S. Chen, "Space Remote Sensing Systems", Academic, Orlando (1985)

29. S. Montán and S. Svanberg, Appl. Phys., B38:241 (1985)

30. S. Montán and S. Svanberg, Industrial applications of laser-induced fluorescence, L.I.A. ICALEO, 47:153 (1985)

31. S. Svanberg, Phys. Scripta, T19:469 (1987)

32. P. S. Andersson, S. Montán and S. Svanberg, IEE J. Quant. Electron., QE-23:1798 (1987)

33. S. Svanberg, Phys. Scripta, T16:90 (1989)

34. S. Andersson-Engels, J. Ankerst, A. Brun, A. Elner, A. Gustafson, J. Johansson, S.-E. Karlsson, D. Killander, E. Kjellén, E. Lindstedt, S. Montán, L. G. Salford, B. Simonsson, U. Stenram, L.-G. Strömblad, K. Svanberg and S. Svanberg, Ber. Bunsenges Phys. Chem., 93:335 (1989)

OPTICAL FIBER SENSORS FOR THE ENVIRONMENT

J. P. Dakin

York Ltd.
Chandlers Ford
Hampshire, U. K.

I. INTRODUCTION

The limited ability of the Earth to cope with the extremes of Man's pollution has been receiving an increasing amount of political attention in recent years, with greatest media emphasis being on global effects, such as the "greenhouse" effect and the well-publicised "holes" in the ozone layer.

These effects would be particularly frightening, if they were to continue unabated to their predicted long-term conclusions. However, there are also many more localised incidents of environmental pollution, for example in home and industrial environments, which currently cause acute problems and may result in cases of severe hazard or death. These are, by their sporadic nature, often difficult to fully assess and quantify. In addition, there are many examples of wastage of valuable resources, such as heat energy loss, chemical spillage etc., which may cause extensive economic loss.

In modern society, the greater capability of technology to determine, more accurately, the level of pollution and wastage is increasing the need for suitable equipment to perform continuous monitoring of the surroundings. The drive for installing sensing equipment is not merely driven by the moral urge to protect Man and the Environment, (indeed, certain cynics may claim this is often far from the case) but is becoming of direct economic concern to companies, due to the increasing number and cost of cases of litigation, in response to pollution hazards. In addition, there is now an increasing level of responsible, preventative legislation, aimed at preventing hazardous incidents.

This article summarizes the potential which optical fiber sensors have for the remote monitoring of the physical and chemical properties of the surroundings. Emphasis is placed on the environmental and health and safety aspects of chemical pollution, with discussion on related factors,

Optoelectronics for Environmental Science, Edited by S. Martellucci and
A.N. Chester, Plenum Press, New York, 1990

such as thermal energy wastage. These areas.are currently receiving significantly more attention, as a greater realization of the Earth's limited ability to cope with the greater excesses of Man's activities is becoming apparent.

The use of optical fiber sensors in this area has, so far, been extremely limited. However, they do have many potential advantages to offer. Their principal attributes are the following:
1. Low attenuation, allowing remote analysis over many km.
2. Capability to perform <u>distributed</u> measurements, again over many km of fiber.
3. Ability to perform a wide variety of measurements via optical fiber.
4. Non-intrusive nature (non-electrical, negligible ignition risk, freedom from lightning damage).
5. Small physical size, even when in armoured form, relative to conventional cables.
6. Highly specific and direct nature of many optical measurements, and high speed of response.
7. Excellent corrosion resistance of optical fibers.
8. Possibility of incorporating self-checking algorithms in optical analysis terminals.
9. Ability to maintain good spatial coherence in long length of mono-mode fiber (when desired for coherent sensors).

Environmental sensing applications may be conveniently classified, according to the type of area to be monitored. The factors determining the applicability of sensors in these scenarios will now be discussed.

II. SENSING IN THE HOME

The home environment is where Man spends most of his time, and where he often has a significant exposure to many natural and man-made hazards. However, this is an extremely cost-sensitive area, and therefore the least easy one to employ the first examples of any new technology. Indeed, many of the safety problems in the home tend to be covered by attention to con-structural design regulations, rather than by using expensive sensors. Very few monitors, of any kind can be produced at the necessary low costs; the only ones, so far, being smoke and gas sensors of low-cost electrical type. Any optical fiber sensor would require to be of extremely low cost, and probably of a multiplexed nature to find a place in the home. It is unlikely, in the near term, that any fiber sensors will meet the severe cost constraints for home pollution monitoring.

III. INDUSTRIAL MONITORING WITHIN BUILDINGS

The internal factory environment is the area in which the most extensive use of physical and chemical sensing is made. However, again, the cost of individual sensors is generally a major factor, and the constraint of, often-irrational, user conservatism can often be an important consideration. However, modern industry is tending to evolve more rapidly than previously, and is making increasing use of a greater variety of automatic monitoring and control systems. These are leading to greatly reduced manufacturing costs and superior quality control.

In the environmental areas, the twin-drive factors, vis increasing preventative safety legislation and punitive damages for accidental leakages, are producing a strong incentive for new monitoring equipment for hazardous materials. In addition, the control of wastage of both material and energy are of economic importance.

Perhaps the two main factors which will fuel the need for fiber sensors for these applications are their high level of intrinsic safety in flammable areas, and the valuable information that can be derived using remote, optical spectroscopic analysis. The methods become more cost effective if the opto-electronics terminal units can be centrally located in a control area, with the passive sensing heads being sequentially polled, using a low-loss optical switch. A typical layout is shown in Figure 1. Clearly, the switch must be capable of good repeatability charac- teristics, as otherwise, some form of alternative referencing method (for

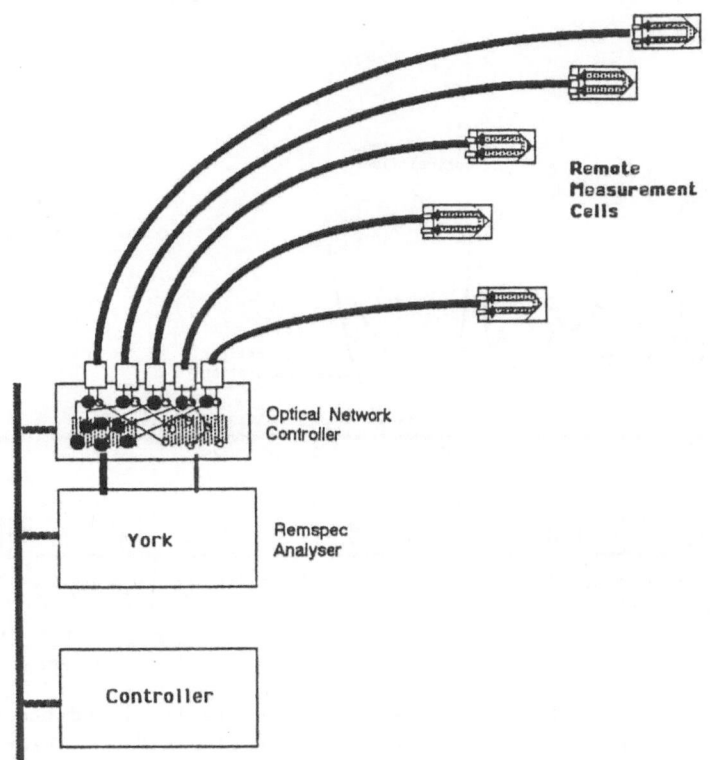

Fig. 1. A schematic illustration of a networked VIS-NIR remote spectroscopy system, using a computer controlled switch (York Switchmaster) and fiber-compatible remote spectro- meter (York Remspec).

example two-wavelength referencing) will be necessary. York has already conducted networking and achieved a reproducibility of less than 0.1 dB for their multimode version of the switch, with the feedback-control optimisation.

The two main applications of switch-multiplexed networks are monitoring of pollution, or "on-line" analysis of chemical process plant, using fiber harnesses with passive optical sensing heads. Many of the traditional spectrophotometer procedures may be produced in optical fiber remoted form, removing the need for batch sampling and subsequent analysis. Thus, with such a system, a wide variety of types of environmental monitoring may be performed, provided the media of interest are capable of spectrophotometric analysis, at the levels required and within the spectral transmission window of the fiber used.

The range of gases that can be monitored using conventional silica optical fibers is shown in Figure 2. This is redrawn from an original paper by Inaba[1], who was one of the first to suggest the use of optical fibers for remote pollution monitoring. Since Inaba's early paper, many more techniques have been devised to improve the sensitivity and stability of gas sensing, with the main emphasis being on flammable gases, because of their explosion hazard[2,3]. The optical detection of gases is usually

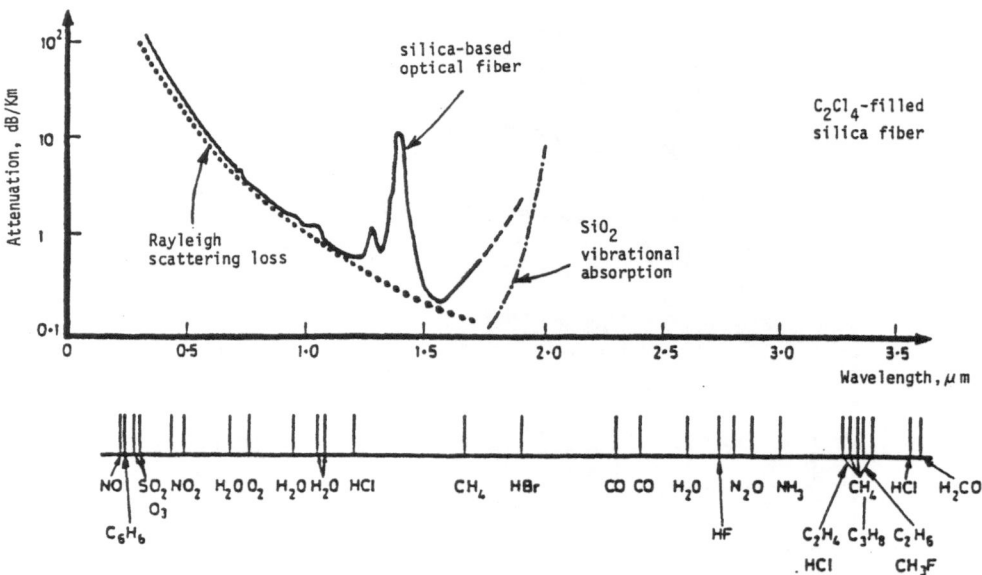

Fig. 2. Attenuation of silica-based fiber (solid line) and the limits imposed by Rayleigh scattering (dotted) and SiO_2 vibrational absorption (dot-dash). Lower scale shows major gas absorption lines of common pollutants. (Diagram is redrawn from ref. 1, by Inaba). Note: More recently developed fluoride fibers extend the measurable range to approximately 4.0 μm in the infra-red region, where absorption lines are much stronger.

achieved via their electronic or vibrational (or rotational), absorption lines, which are strongest in the ultra-violet and infra-red regions respectively. Thus, the availability of special fibers, with enhanced UV or IR transmission, is likely to expand the range of gases which may be visibly detected.

Many infra-red fibers (such as fluoride glass types, with transmission out to beyond 4 μm) are already available[4] . However, most commercially-available types appear to have a rather high cost and there are still doubts regarding their mechanical and chemical durability, compared to the well-established silica-based type. Ultra-violet transmitting fibers, with either high OH content silica, or with liquid-filled cores have been available for some time. The use of distributed sensors for monitoring industrial environments is lileky to occur in two main areas. The first is that of distributed temperature sensing, which can form the basis of a wide variety of sensors for process monitoring and for estimation of thermal efficiency (or, conversely, the degree of energy wastage). In addition, there are sensors capable of distributed sensing of polluting chemicals. The latter pollutants may be gases, capable of modifying the evanescent field propagation in a reactive fiber cladding[3,5], or may be radioactive materials, which may more directly modify the absorption loss[6] of the fiber core material, by means of gamma-ray or neutron interactions[6]. In most distributed sensors, some form of Rayleigh scattering[7,8] or Raman-scattering[9,10] optical-time-domain reflectometer will be utilized. The first distributed sensor to be commercially available is the Raman distributed temperature sensor (Figure 3). (Diagram taken from an early publication of the basic method). This system is being considered for a wide variety of thermal monitoring applications in industrial environments. The instrument has significant potential for monitoring (and hence enhancing) the thermal efficiency of industrial plant.

In addition to transmission monitoring, an important aspect of fiber sensing, which has not yet been mentioned, is that of particle sensing. There are many conventional optical particle sensing methods, based on scattering of focussed optical beams, which may find counterpart in fiber-remoted form. The methods, likely to be most applicable to airborne particle sensing, are those in which scattering from individual particles is monitored, as they pass through a narrow focussed beam. The use of such systems for the characterisation of smoke and dust is an important application. (In addition, the use for water purity monitoring will be considered later - See Section 5).

If it is desired to determine the size of particles from the scattered beam, it is essential to ensure that only light from particles, which are well centred in the beam, is monitored. Otherwise, particles at the beam edges, which receive a lower intensity of incident light, will have a smaller apparent size, as determined from the observed scattering cross-section. The solution to this is to arrange for a "trigger" beam, concentrically located within the central region of the main beam. When light is scattered from this trigger beam, the small scattering particle must be centrally located. A novel solution to this problem is shown in Figure 4, where a graded index lens, with inherent chromatic dispersion, produces

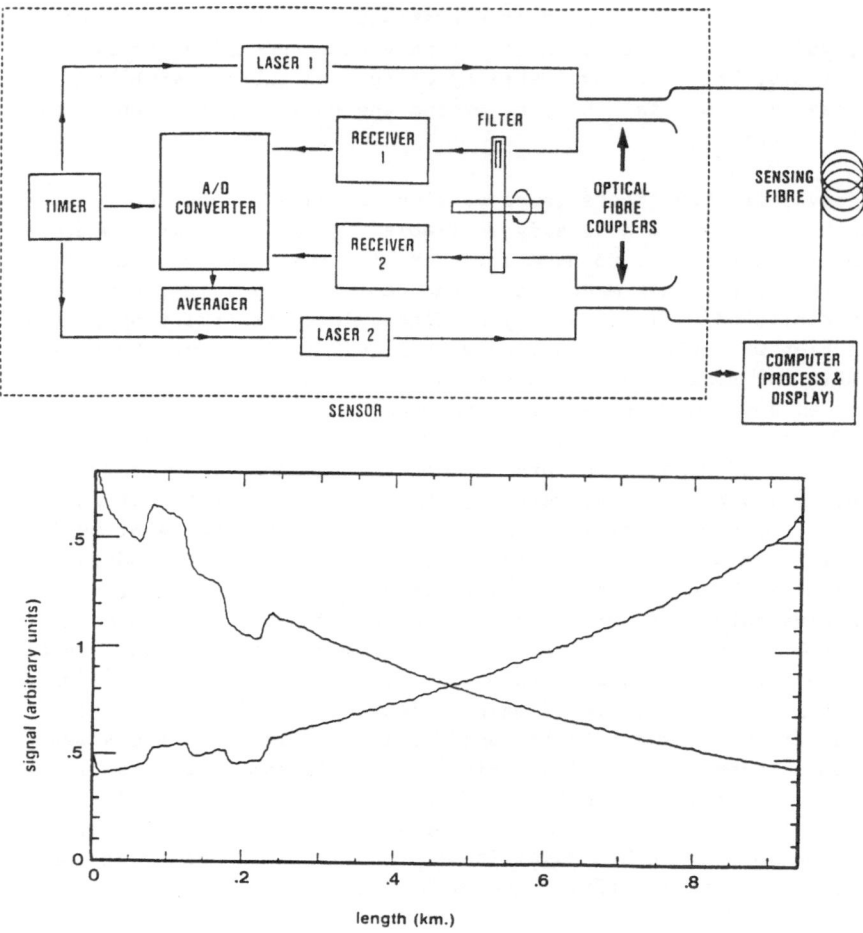

Fig. 3. Schematic diagram of an early version of a commercial distributed temperature sensor system[10]. The lower curve shows the measurement of Raman anti-Stokes signals from opposite ends of a fiber, to allow correction for fiber variations. (More recent versions have no moving parts, but allow multiplexing of measurement fibers).

longitudinally-displaced beam waists, with the desired diameter differences and guaranteed concentricity. The scattering volume is defined by the overlap of the incident beam and the collection zone of the ortho-gonally-situated collection optics, which again use a similar fiber/graded-index-lens arrangement. (This method has been devised by a York collaboration with the U.K. public health laboratories and RSRE in care of R. Carr and E. Perkins).

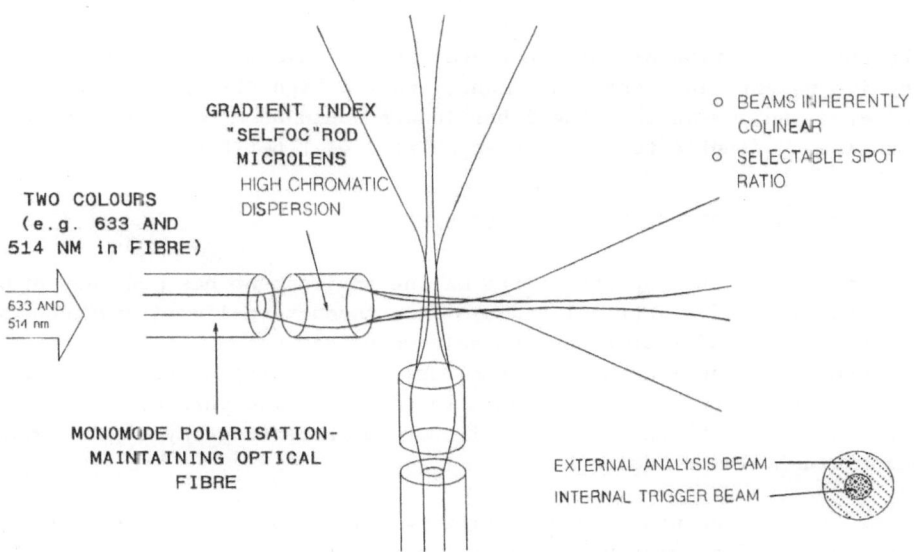

Fig. 4. Two colour, concentric beam optical fibre
for single particle detection.

IV. OPEN-AIR AND SPACE SENSING

It is in open-air sensing that the low loss of optical fibers may be
used to best effect, to achieve long-range measurements over wide areas.
In the spectral region between 1.0 μm and 1.7 μm, the attenuation of
silica fibers lies mostly below 1.0 dB/km and may drop to as low as 0.2
dB/km in the 1.45 - 1.6 μm region. Thus, it is possible to carry out
spectroscopic measurements on materials, which absorb in the low-loss
window of the fiber, at distances up to 10 km, or so, if it were desired.

The main applications for open air gas sensors are currently for
methane sensing on oil and gas drilling rigs[11]. However, there is potential demand for control of emission from chemical plants and from exhaust
chimney stacks. The main differences, when compared to indoor monitoring,
are the greater fiber cable distances that may be involved, and the lower
concentrations of gas that will usually be present in the dispersed
outdoor environment. As with in-factory monitoring, the capability to
multiplex a number of passive networks, using a stable, low-loss switch,
is likely to be of major advantage. The development of distributed
sensors, using reactive claddings, may be an attractive proposition for
the simultaneous monitoring of hazardous leaks of higher concentration.

The use of fiber optics for extra-terrestrial sensing of environmental aspects is still a rather speculative concept. However, the detection of ionizing radiation, using long fiber probe lines, could be an
attractive research tool. One advantage would be the long length of fiber
that could be deployed in weightless conditions. It could, for example, be
envisaged to use fiber umbilicals, suspended from orbiting spacecraft, to

probe the composition of the outer atmospheric layers of planets. The low effective weight, in these conditions, and the high strength/weight ratios of fibers, would mean that the fiber length would be more likely to be determined by fiber attenuation than by risk of fiber breakage.

V. MARINE AND WATER SUPPLY PURITY SENSING

The use of fiber optics in the marine environment has many advantages, due to the lack of electrical signals in a conducting-liquid environment and the high immunity of fibers to salt-water corrosion. In addition, the small cross-sectional area can reduce the viscous drag of towed cables. The use of fiber probes may potentially detect a wide variety of polluting conditions, or, with the use of deliberate markers (e.g. dyes), be used to examine tidal flow patterns[12].

As before, the use of a multiplexing switch can ensure the most cost-effective use of a relatively sophisticated measurement station, and again, the use of distributed sensors could have attractions, if suitable fiber claddings can be developed.

One major difference between gaseous and aqueous spectral absorption lines, is that the former are often very narrow, whereas the latter are usually broadened significantly, due to the field interactions arising from the close proximity of solvent molecules. Thus, only species causing a relatively major change in the absorption level will usually be detectable in the case of marine sensing.

However, the aqueous environment is particularly well suited for scattering sensors designed to monitor suspended particulate matter, as the flotation aspects of this denser medium and the stronger influence of molecular collisions, greatly favour production of stable turbid suspensions. Thus, aqueous suspensions more readily permit the use of light-scattering techniques, such as photon-correlation spectroscopy, using optical fiber links[13], (Figure 5). This method for particle size analysis involves illumination of a small scattering volume, and the observation of scattered light from the volume. The changes in scattered intensity, due to interference of scattered light from many individual particles, each moving relative to its neighbours, as a result of Brownian motion, allows a measure of the particle size to be calculated. For this method to give accurate results, it is necessary to have a sufficiently high number of particles in the sample volume, to ensure that scattered intensity changes occur primarily due to particle motion, rather than as a result of changes in the number of particles in the scattering volume.

VI. CONCLUSIONS

This paper has attempted to set the scene, by describing the application areas in which fiber optic environmental sensors might be used. The suggested references will describe the examples given in more detail, and will present examples of further sensor types, to expand on the treatment above.

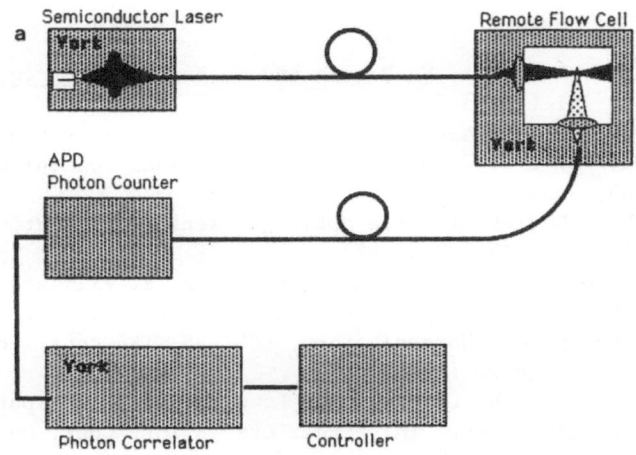

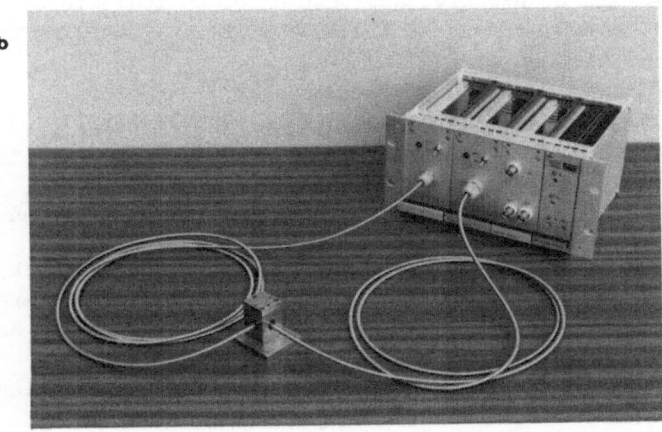

Fig. 5. (a) Schematic diagram of a photon correlation
spectroscopy apparatus using remote fibre-optic
probes; (b) York prototipe photon correlation
spectrometer with semi-conductor source and
detector plus optical fibre leads.

In spite of the relatively early stage of development of many optical
fiber sensors for use in this area, there would appear to be significant
research potential, and, provided the systems are well engineered, they
may offer significant advantages over more conventional solutions. The
economic success of fiber-based systems is likely, in many cases, to
depend on the viability of multiplexed or distributed approaches to
measurement problems, and the reliability of the optical solutions, in
application areas where the probes may often be subject to surface
contamination.

REFERENCES

1. H. Inaba, T. Kobayasi, M. Hirama and M. Hamza, <u>Electron. Letts.</u>, 15, 749-751 (1979)

2. A. Hordvik, A. Berg and D. Thingbo, A fibre optic gas detection system, <u>in</u>: Proc. 9th Int. Conf. on Optical Communications, ECOC 83, Geneva (1983)

3. J. P. Dakin, Review of fibre optic gas sensors, <u>in</u>: Proc. SPIE, vol. 1, 011, 173-182 (1988)

4. P. W. France, S. F. Carter, M. W. Moore, J. R. Williams and C. R. Day, Status and prospects of mid I-R fibres, <u>in</u>: "Proc. 14 th Int. Conf. on Optical Communication, ECOC 88, IEE conf. pub. no. 292, 428-432, Brightom (1988),

5. R. A. Liebermann, R. L. Blyer and L. G. Cohen, Distributed fluorescence oxygen sensor, <u>in</u>: "Proc. OFS 88 Int. Conf. on Optical Fibre Sensors", 346-348, New Orleans (1988)

6. W. Gaebler and D. Braunig, Application of optical fibre waveguides in radiation dosimetry, <u>in</u>: "Proc. OFS 83 Int. Conf. on Optical Fibre Sensors", London (1983)

7. J. P. Dakin, <u>J. Phys.</u> E 20, 954-967 (1987)

8. A. H. Hartog, <u>J. Lightwave Tech.</u> LT-1, 498-509 (1983)

9. J. P. Dakin, D. J. Pratt, G. W. Bibby and J. N. Ross, <u>Electron. Letts.</u> 21, 569-570 (1985)

10. A. H. Hartog, A. P. Leach and M. P. Gold, <u>Electron. Letts.</u> 21, 1061-1063 (1985)

11. S. Stueflotten, T. Christensen, S. Iversen, J. O. Mellvik, K. Almas, T. Wien and A. Graay, An infra-red fibre optic gas detection system, <u>in</u>: Proc. OFS 84 Int. Conf. on Optical Fibre Sensors, 87-90, Stuttgart (1984)

12. J. P. Dakin and A. J. King, Limitations of a single optical fibre fluorometer system due to background fluorescence, <u>in</u>: "Proc. OFS 83 1st Int. Conf. on Optical Fibre Sensors", IEE conf. pub. no. 221, 195-199, London (1983)

13. K. M. Chow, A. G. Stansfield, R. J. G. Carr, S. G. Rariry and R. G. W. Brown, <u>J. Phys.</u> E 21, 1186-1190 (1988)

FIBRE OPTIC CHEMICAL AND BIOCHEMICAL SENSORS

FOR ENVIRONMENTAL MONITORING

A. L. Harmer

Harmer Associates
Geneva, Switzerland

I. INTRODUCTION

This lecture covers the principles of optical fibre sensors for
chemical and biochemical measurements applied to environmental sciences.
Optical techniques and optical spectrometry are widely used in environmental monitoring with classical optical systems, and their adaption to
optical fibre technology further enhances their capabilities. Fibre optics
allows the use of remote probes (up to 5 km), for site monitoring over a
large area or for deep measurements in lake and sea water. Fibre probes are
small and occupy a low volume. Thus, they can be used in inaccessible
places (e.g. ground water monitoring, down-hole sensors). They are inert
and non-electrical and are suitable for hazardous areas, such as petro-
chemical plants, to avoid fire and sparks risk. They are also highly
sensitive, compact with economy of reagents for low cost. They can be used
as disposable sensors or permanently-deployed sensors.

Disadvantages include sensitivity to ambient light, photodegradation
of the chemical reagents by the exciting light source, drift due to
reagent instabilities, and non-reversibility. The disadvantages may be
overcome with further development.

II. GENERAL PRINCIPLES

Fibre optical chemical sensors generally fall into three categories:

i) Fibre acting as a simple light guide carrying light to a probe (e.g.
    to excite fluorescence) and a return fibre to bring the fluorescent
    light back to the photodetector. Absorption measurements normally
    require a through path, or folded by adding a small mirror. The
    volume sensed by the light depends on the probe geometry. Selection
    of a suitable source and detector depends on the application: wave-
    length range of operations, required spectral line width, required

incident power, losses occurring in the optical transmission and the sensitivity of the photodetector. Important limitations are imposed by the attenuation of the fibre: silica fibre has three orders of magnitude higher attenuation in the UV (200 dB/km) compared to the low loss window at 1.5 μm (0.2 dB/km). New fibres, such as fluoride-based types, can extend transmission to longer wavelengths (2-10 μm) for detection of chemical organic molecules using the "fingerprint region" of vibrational absorptions.

ii) <u>Surface waveguide (evanescent mode devices)</u>. A declad fibre or integrated optic slab waveguide allows penetra-tion of the light from the core into the surrounding surface layer by the evanescent wave. This is a sensitive method of monitoring chemical reactions in the surface layer: e.g. for refractive index, 2-phase flow, and chemical and biochemical reactions occurring on a surface.

iii) <u>Chemical reagent sensor</u>. The sensor head uses a chemical reagent surrounding the fibre tip, and enclosed by a permeable membrane separating the reagent from the surrounding medium and providing selectivity for the species diffusing through the membrane. This type of device is termed an "optode" or "optrode".

Special spectrometers are required for fibre coupled sensors with appropriate optics to match the fibre numerical aperture and the small core size. This allows miniaturization of the spectrometer, integration with the processing electronics, and the possibility of multiplexing several fibre-coupled sensing heads into a single centralized instrument.

III. APPLICATIONS

Applications are widespread for fibre optic chemical sensors. They include measurements of absorption, fluorescence and fluorescence quenching. Absorption measurements may be made by a straight through path or by diffused reflection with light being returned from the medium after scattering. Fluorescent quenching is an extremely sensitive technique for specific reactions. Work to date includes the detection of heavy mass cations (uranium, plutonium, etc.) and light metal ions (beryllium, sodium and many others); anion concentrations (chlorine, iodine and other halides); gases and dissolved gases (H_2O_2, ammonia, HCN. etc.); pH; oxygen and CO_2 concentration (PO_2, PCO_2); biomedical parameters (immunoassays, glucose concentration); hydrocarbon gases (methane, ethane, etc.); moisture and humidity; phosphate esters; and many other chemical compounds. Applications can be found in measurements of liquid levels, refractive index, absorption, fluorescence and colour spectrometry, scattering (nephelometry and turbidity).

IV. GAS SPECTROSCOPY

Fibre absorption cells have been used to detect gas molecules, mainly explosive gases such as methane and ethane, but also CO, and NO_2. To achieve a good sensitivity, a long cavity length is required (typically 50 cm), and the contrast can be enhanced using multiple pass cells. Differential absorption techniques improve the stability by measuring the required wavelength and a reference wavelength simultaneously. Measurements up to 5 km distance have been made with a resolution of 0.25% CH_4.

Raman spectroscopy has been made on gas flames measuring N_2 in a CH_4-air flame.

V. REFRACTIVE INDEX

Refractive index. two-phase flow mixtures, and liquid level have been determined from a measurement of the critical angle of refraction at the fibre-liquid interface. A large number of probe-geometries are possible (prisms, unclad fibres, bent fibres, etc.), and sensitivities of 10 better than 10^{-5} refractive index units achieved. These probes can be used for online industrial measurements, or environmental monitoring of small index changes.

VI. TURBIDITY (OR SCATTERING) MEASUREMENTS

Fibre sensors have been developed for oil-in-water monitoring for measuring the oil pollution in effluent water streams, or discharge from tankers. For marine use the required range for ballast dumping is 1000 ppm. For inland waters the oil content is below 0.2 ppm ± 0.1 ppm. The sensor measures forward and sideways-scattered light, and computes the scattering content.

VII. pH (HYDROGEN ION) CONCENTRATION

A large number of pH sensors have been developed based on colourometric reactions of standard dye systems with different membrane materials. The sensor uses the pH colour reagent immobilized onto a substrate (polymer micospheres or a membrane) and the light uses a dual wavelength system to distinguish colour changes at the peak absorption of the indicator dye and a second reference wavelength which is insensitive to pH changes. Most pH sensors have been directed towards medical applications (pH range 6.5 to 7.5) but some experiments have been used for environmental monitoring.

VIII. OXYGEN AND CARBON DIOXIDE CONCENTRATION

Fibre optic sensors for oxygen concentration are based on fluorescence quenching, with a fluorescence dye absorbed on a plymeric substrate membrane which is oxygen-permeable. Absorption of oxygen through the membrane reduces the fluorescence efficiency of the dye. No chemical reaction is involved: the fluorescence quenching occurs as a result of an electronic energy transfer from the photoexcited molecule to molecular oxygen. Thus the response is both rapid and reversible. A number of dye systems and suitable membranes have been examined, with different forms of encapsulation such as microbeads and a range of membrane materials.

CO_2 concentration can be measured by monitoring the pH of a bicarbonate/carbonic acid equilibrium mixture according to the mass-action equation. The sensor consists of a CO_2-permeable membrane enclosing a bicarbonate solution in a buffer gel. A pH indicator is added with a suitable pH value, and the bicarbonate concentration of the indicator buffer gel is adjusted to cover the appropriate range. Most work on oxygen and CO_2 sensing have been directed towards blood gas monitoring for medical purposes, but can be readily extended to industrial measurements.

IX. CHEMICAL ION CONCENTRATION

Monitoring techniques using remote fibre optic fluorescent measurements have been applied to the remote monitoring of numerous chemical species, such as UO_2^{++}, actinides, sulphates, chloride and halide ion concentrations, H_2S, etc. Main applications are in pollution monitoring and detection of trace impurities in ground and surface water. The usual light source is a high powered ion laser and the fluorescent light is collected by a return fibre, filtered with a monochromator and detected using photon counting techniques. Different probe geometries have been developed: a simple fibre end, a ball-shaped lens and semi-permeable membranes, etc. For uranium ion detection sensitivities of better than 10^{-14} molar can be achieved by enhancing the sensitivity using phosphate co-precipitation.

X. EVANESCENT WAVE SPECTROSCOPY

Sensitive surface chemical reactions can be detected using evanescent wave spectroscopy. When the naked core of a waveguide is placed in a liquid sample, the evanescent wave of the light guided by the waveguide penetrates into the surrounding liquid for a distance of the same order of the wavelength of the light and can measure reactions taking place on or very close to the surface of the guide. The intensity of the evanescent wave may be increased by plasmon resonance which uses a thin metal film on the surface and the light induces a collective oscillation in the free electron plasma at the metal-dielectric boundary. This technique has been used for a variety of applications: biomedical reactions such as immunoassay, gas sensing by coating the waveguide with a gas-sensitive membrane which swells as a result of gas infiltration, and other chemical species. Sensitivities of 10^{-10} molar concentrations or better are possible.

XI. DISCUSSION

Optical fibre technology offers significant advantages over conventional chemical sensor systems, the major advantage being that the energy source and information processing may be separated by a long distance from the sensor using optical signals. A wide variety of approaches is possible. However many problems are still to be overcome, particularly for long-term stability resulting principally from poor stability of many of the sensing reagents. In addition, there is often a slow response time arising from the need for mass transfer from the analyte to the reagent which are generally in different phases. Future development concerns the mechanical design of the probe-to-fibre coupling, the reproducibility of the optical cell especially when miniaturized probes are required. With improved development and adaption to the specific working conditions, chemical fibre optic sensors offer considerable future potential for environmental monitoring applications.

REFERENCES

1. See the collection of articles in a single volume: Talanta, Pergamon Journals Ltd, Vol. 35, No. 2, (1988)
2. A. L. Harmer and R. Narayanaswamy, Spectroscopy and Fibre Optic Transducers, in: "Chemical Sensors", T. E. Edmonds, ed., 275-294, Blackie and Son, Glasgow, (1987)
3. C. Nylander, J. Phys. E Sci. Instrum., 18, 736-750 (1985)
4. W. R. Seitz, Anal. Chem., 56, 16A-33A (1984)
5. T. Hirschfield, J. B. Callis and B. R. Kowalski, Science, 226, 312-318 (1984)

REALIZATIONS AND PROGRAMS OF ENEA IN THE FIELD OF OPTOELECTRONICS

M. A. Biancifiori and M. Madaro

ENEA*, TIB/SCIENT
CRE Casaccia
Rome, Italy

I. INTRODUCTION

Optic and electro-optic technologies are being developed in all the industrialized countries, because they are one of the most interesting technological fields from both a scientific and an economical standpoint. They have a number of different applications, including many of great strategical importance, such as: informatics, robotics, microelectronics, telecommunications, advanced materials, mechanical processing (welding, cutting, drilling, surface treatments, non-metallic materials working), industrial and environmental control and biochemistry.

In Italy, a number of laboratories in industries, universities and research centres are involved in the development of new lasers or the improvement of the existing ones. They are performing a broad spectrum of research and development activities, both on the theoretical and on the experimental side, ranging from the development of laser sources and optic and electro-optic components, to the realization of complete systems for photochemistry, teledetection and material processing.

ENEA is one of the leading institutions in laser R&D since the early 60's, when the first laser systems were developed for the production and diagnostics of plasma of thermonuclear concern. Complex optic systems were realized using a ruby laser[1], where a light emission at the wavelength of 700 nm was obtained by the optical excitement of a synthetic ruby crystal of particular properties. The peculiar features of lasers, such as the possibility to concentrate energy on very small areas, suggested to use lasers directly for the production of ionized gases at high temperature, by focusing a high power laser beam on a suitable target, in order to prime the thermonuclear fusion[2]. In the years from '64 to '70, a ruby laser (550 MW power for 20 nsec) and a neodynium laser (3000 MW power

---

*ENEA stands for the Italian National Committee for Research and Development of Nuclear and Alternative Energies.

for 10 psec) were realized[3]. Together with laser sources development, R&D activities were also started on advanced optic and electro-optic systems for photochemistry, molecular spectroscopy and isotope separation: many of these activities are still continuing.

ENEA is a public agency, performing research, development and technology transfer activities in the fields of energy, environment and innovative technologies; it runs a number of projects intending to provide public support to the national industry, in order to make it competitive in the international market. These projects put together capabilities existing in ENEA itself, as well as in universities and industries, in order to achieve scientific and technical realizations, that can be significant not only for the participants, but also for the whole Italian economy. At present, ENEA is formed by four areas (Innovative Technologies, Environment, Energetics, Nuclear Energy): the laser-related activities are performed mainly in the area "Innovative Technologies", while many environmental and biological applications of optoelectronics are carried out in the area "Environment".

II. ENEA PROJECT ON "OPTICAL AND ELECTRO-OPTICAL TECNOLOGIES"

The ENEA project "Optical and electro-optical technologies" is, together with the CNR Finalized Research Project on High Power Lasers (from many sides the two programs complement each other), the most important Italian program for the technological innovation in the field of optics. Its purpose is to develop an organic plan in optics and electro-optics, in order to support the evolution of capabilities and their transfer to the industry. The project covers three areas: Optical and electro-optical integrated components; Optical systems; and, Laser sources. A remarkable theoretical work (quantum optics, charged particles beam dynamics, models related to pumping processes of gas laser) is also performed in the project. As far as funding and human resources in the 5-year plan 1985-1989, 35000 millions Italian lire and 400 man-year have been allocated, on the whole.

II.1. <u>Optical and Electro-Optical Components</u>

The overall purposes of the project area "Components" are mainly knowledge acquisition, technological development and engineering of systems and/or processes, in those research sectors considered necessary to be explored for realizing a national reference point in the field of the methodologies of optical and electro-optical devices production. The innovative technologies that appear to be promising with respect to the continuous, reliable and flexible production of a whole class of components and for the applications in related fields are developed. Not only the development of know-how, but also the realization of systems at such an engineering level to make an immediate transfer in the industrial field possible are the purposes of the project. The main realizations in this field have been:
- a dual ion beam sputtering system[4], for the production of multi-layer optical devices with high adhesion and high laser damage threshold

features, using non-traditional materials; by means of such a system, silicon oxide and aluminum oxide films (300-400 nm in size) and diamond-like carbon films have been deposited. The latter have very good mechanical (hardness and adhesion) and optical (in IR) properties;
- a computerized spectrophotometric system for the on-line monitoring of thin film devices during the deposition[5];
- interferometric systems for the control of reflective and refractive optical components during the processing phases.

A support to national industry has been provided by means of R&D in the fields of refractive and reflective optics, with protective and multi-layer coatings: in collaboration with national industry both metal reflective optics for high mean power CO_2 lasers and advanced mechanical components for mechanical-optical processing have been realized. A wider initiative, involving two Italian firms, foresees to realize, by using the ENEA capabilities, a new producing structure to introduce high-quality components in the national market. A particular attention has been paid to the turning by diamond tool, that allows the accurate working of optical surfaces of various shapes to be made, with a surface finishing similar to that obtained by lapping.

Another activity is developing the production of optical materials of traditional type (but not easily available in Italy) or endowed with new physical features, making them interesting for optical applications. The goal of this activity is the development of computerized optic ellipsometry systems and PDS (photothermal deflection spectroscopy) systems, for the determination of optical parameters (refractive index, absorption and thickness) of thin films[6]. Moreover, highly reliable measurement methodologies are developed, in order to realize systems directed to a routinary use during production or for the material classification.

Finally, the application of the results obtained by the cited activities is foreseen, for the production and the testing of prototypes or small series of components, whose demand has no adequate answer on the market. Examples include:
- the production of optical multi-layer devices with a high laser damage threshold, metallic optics (both reflective and diffractive) for IR and visible, processed by diamond tool, refractive optics for IR, variable reflectance mirrors (VRM);
- components control and certification systems (that could also be used in the perspective of the realization of national technical-scientific services by ENEA), such as: spectrometers for reflectance absolute measurements in visible, UV and IR, systems for the determination of surface and volume features, high UV fluence damage testing systems, computerized interferometric systems for the macroscopic control of reflective and refractive components in IR.

II.2. Optical Systems

The area "Optical systems" has the goal of realizing highly innovative electro-optic systems and lasers (not developed at present by other Italian institutions). Three major kinds of activities have been given the priority:

a) the advanced photochemical laser systems for the production of strategical materials for mechanical and electronic industry, for the production of solar energy and for the control of energy production and combustion processes;

b) advanced electro-optic and optic-mechanical systems for the industrial applications, taking into account the remarkable help that they can provide in the definition and control of production processes, as well as for their application in the research field;

c) systems for the remote detection of both natural and energy-connected pollution by means of lasers (LIDAR, DIAL), and for terrametry, anemometry and laser thermometry.

The monochromaticity of laser light allows to induce photochemical processes on gas phase molecules or at the interface gas/surface. By means of UV or IR lasers, it is possible to selectively break only some chemical bonds, driving the successive reactions towards the desired final products. Among the photochemical processes considered in the project we find the production of powders to be sinterized and the deposition of high-quality thin film coatings. The realization of photochemical reactors for the production of sinterizable powders and for the deposition of thin films is one of the main purposes of the project. Powders of silicon carbide and nitride, catalytic powders (e.g. TiO_2), powders containing heavy metals (WC, CrC_2) for anti-corrosive and anti-wear films can be prepared[7,8]; powders of pre-fixed dimensions and high uniformity can be obtained only by laser dissociation of one or more compounds in gas phase. The production of such powders is of great interest for mechanical and car industry. The laser induced synthesis of ceramic powders allows to obtain products with such features to optimize the quality of the produced ceramic materials (as the resistance to high stress levels, at high temperature too). SiC and Si_3N_4 powders can be used for ceramic materials sinterization or for high quality surfacing and alloying of steels. The ceramic materials quality depends upon the properties of the powders used for sinterization: the features of ideal powders are dimension uniformity, diameter $< 1\mu m$, high purity. All these requirements can be fulfilled producing the powders by means of chemical reactions among the suitable gas precursors, primed by a CO_2 laser.

Deposition processes of anti-corrosive and anti-wear films (Ti, TiN_x, or doped amorphous silicon,...) have applications in microelectronics (litography, metallic interconnections in integrated circuits, ohmic contacts in semiconductors, metal or insulating layers deposition on integrated circuits) and in solar energy (silicon photovoltaic cells).

Thin films of metals, insulators and semiconductors can be deposited by laser dissociation of a suitable mix in gas phase. This method presents the following advantages with respect to the conventional ones:

- low substrate temperature
- localization of reaction volume, and
- absence of radiation damage.

Silicon nitride is a very interesting compound for microelectronics, because of its impermeability to moisture and alkaline ions and its high dielectric constant. Its deposition can be obtained by an excimer laser (ArF, 193 nm) from a mix of ammonia, silane and nitrogen. The deposition of zinc is important for many applications, ranging from the doping of GaAs and InP to the growth of ZnO and semiconducting binary compounds. Zinc thin films can be obtained by the excimer laser dissociation of its alkyls[9]; the gas phase photoreactions can be investigated by using the technique of multiphoton ionization, coupled to time-of-flight mass spectrometry[10].

The excimer lasers have been applied also in the decomposition of pollutants such as polychlorinated biphenyls, that are considered precursors of dioxine compounds[11]. The tunable lasers, moreover, are very important sources for spectroscopy, as well as for isotope enrichment. Several activities related to the development of advanced spectroscopic systems, such as very high resolution spectroscopy, Coherent Anti-Stokes Raman Scattering (CARS), Laser-Induced Fluorescence (LIF), REsonant Multi-Photon Ionization (REMPI), are performed. These systems are needed for temperature determination and for reagents and products control (including transitory ones such as radicals), in order to optimize photochemical processes, as well as for monitoring "in situ" combustion gases composition and temperatures, pollutants concentrations, etc. in limited volumes but with good spatial resolution. CARS[12] has, with respect to other optical diagnostic techniques, the advantages of a high conversion efficiency, a simple optic line for the detection of the emitted laser beam and the automatic discrimination from luminescence and spontaneous fluorescence phenomena. A high applicability (almost all the molecular species) and a high resolution both in space and in time are the other features of this technique. Temperature and concentrations of biatomic molecules (N_2, H_2, O_2, CO, NO, HF, HCl) and small polyatomic molecules (CO_2, H_2O, NO_2, NH_3, PH_3, AsH_3, CH_4, C_2H_2, C_2H_4, C_2H_6) can be measured on-line, also in presence of large excess of other gases. As to molecular spectroscopy, another goal is to realize, by means of laser diodes, a measurement system of very high resolution molecular absorptions and small gas traces in the atmosphere. The system can be used for photochemistry, environmental pollution and non-invasive medical diagnostics. The activities performed in this field involve:
- development and characterization of spectroscopic systems stabilized in frequency and amplitude;
- determination of molecular parameters of several gases of photochemical and environmental interest;
- feasibility study of spectroscopic measurements for medical applications[13,14].

Activities of infrared multiphoton spectroscopy on polyatomic molecules (SF_6, SiF_4, freon,....) are performed, studying the absorption of IR radiation by molecules cooled in a molecular beam expanded at supersonic speed. The sequential absorption of more than one photon by the molecules in the jet is detected by a bolometer put in the jet itself. The variations in the power transported by the beam, produced by the irradiation with a pulsed IR laser, are estimated to be of the magnitude of 10-100 nW and are detected by the bolometer for every single pulse. Systems for rare isotopes enrichment of weighable amounts of medical, nuclear and geophysical interest (^{33}S, ^{13}C, ^{14}C, ^{2}H, ^{3}H, etc) are also being developed. Tunable narrow-band IR-UV-VIS sources, with good beam quality are needed, as well as middle size process reactors containing supersonic beams for the recovery of the isotopes of interest.

Finally, a high mean power (up to 2 kW) CO_2 laser source is used to perform laboratory tests on cutting structural materials and to study the physico-chemical properties of the produced aerosols: this work is performed within an European program for the dismantling of nuclear power plants.

Optical methods presents several advantages with respect to the traditional ones, such as high accuracy, non-intrusivity, possibility of optical data pre-processing. This way, their use allows a real technological innovation in the interested fields. Particularly, automated interferometric and holographic systems for optical metrology and non-destructive defect analysis are being developed, in order to be able to operate in situ plant inspection. For a better utilization of holographic methods, already employed by industries for non-destructive testing, the development of suitable data registration and elaboration techniques (such as expert systems or image analysis systems) and calculation methodologies are needed. For this reason, the activities related to holographic techniques will be addressed to both optical and mechanical planning and image analysis and data elaboration. Micromovement systems for hostile or non-conventional environment (in vacuum, in presence of strong magnetic fields or aggressive agents) are developed, in view of their growing demand in several industrial fields. Electro-optic sensors for real-time remote detection of the configuration and movements of autonomous robots are developed, in connection with an ENEA project on robotics. A precision in position of 10^{-2} cm and in speed measurements of some cm/sec, for an average distance of 10 m and an average speed of the magnitude of m/sec, are typically required. Moreover, these systems have to work in hostile environments. In the same frame, photo-electric sensors are developed for speedy image analysis, as well as optical communication channels on middle range distances for autonomous robots. Holographic systems using continuous wave (HeNe, Ar) and pulsed (ruby) laser sources have been realized for optical metrology, structural analysis and non-destructive testing. In situ structural analyses on archaeological findings (the Bronze Horse from the Capitoline Museums in Rome)[15] and on mechanical components, optical components quality analysis and behaviour analysis of the active medium of an excimer laser[16] have been performed by holographic interferometry, using a double pulsed ruby laser.

The modern remote sensing techniques and the connected disciplines (including image treatment, stochastic processes, experimental data elaboration) are essential instruments for ENEA to accomplish its tasks, together with its effort to promote the diffusion of innovative technologies in the national producing system. The two main directives of the ENEA action are: a) to possess capabilities and technical equipment, in order to be able to provide experimental facilities and scientific service to external users; examples are the agreements with public agencies to implement an

observation net for the marine environment quality and the pollution
surveillance of Italian coasts (in collaboration with the Ministry for the
Merchant Navy), or to study the river Po (with the support of Lombardia
regional government);
b) to establish collaborations with other operators to qualify and diffuse
new products and services.

The benefits deriving from remote sensing regard environment in its
globality and in its many aspects, with important contributes to the
solution of many problems, such as: sea and air pollution, long range
climatology, ice monitoring, desertification, localization and inventory
of agricultural and fishing resources. Such systems can be used for:
- the monitoring and the quantitative measurement of atmospheric
pollutants and minor components or tracking of effluents coming from
factories, thermo-electric power plants, oil refineries, pipelines;
- sea bathymetry, measurements of oil release in the sea, pollution
mapping;
- measurements of temperature, pressure and aqueous vapour in the atmosphere;
- studies on atmosphere dynamics (wind speed, gas transport, turbolence)
for meteorologic purposes or air safety;
- measurements of trace components in the atmosphere (ozone, water) for
the study of the effects of both natural and artificial processes on the
biosphere; measurements of distribution, dimensions and chemical composition of stratospheric aerosols for studies on radioactivity balance, ozone
depletion and the cycle of SO_2 clusters causing acid rains;
- detection of terrestrial crust movements for tectonics studies in
seismic and vulcanic areas;.
- military applications of lasers (laser rangefinders, satellite tracing,
imaging,...); and,
- systems for laser transmission on large distances with diode or CO_2
lasers sources, presenting several advantages with respect to the RF
systems now used.

For remote sensing of pollutants four major kinds of systems are
being developed in the ENEA project:
1) A ground based infrared LIDAR/DIAL station, using a pulsed CO_2 laser
source, has been projected and built and will be operating in a short time
in the ENEA Research Center of Frascati[17]; this system uses tunable and
high beam quality laser sources of original design (SFUR, Self Filtering
Unstable Resonator)[18,19]. This system (1 J, 100 nsec, 1 Hz) will be used
for in situ measurement campaigns for the quantitative measurement of
freon, NO, NO_2, SO_2, CO, CO_2, O_3, C_2H_4, toxic gases, etc. in the concentration range 10-100 ppb, with a longitudinal resolution of 15 m and for
distances up to 3 km; it will also be used as a test for deriving the
technical specifications of CO_2 LIDAR to be transported on aircraft or
motor-vehicle.
2) Coherent lidar systems are also being developed for wind velocity
measurements via the Doppler shift technique: they will provide a very
useful tool for monitoring such harmful phenomena as wind shear,
expecially during aircrafts take off and landing operations[20].
3) UV-VIS lidar systems using excimer high-power laser sources with Raman

cells to produce convenient output frequency shifts, suitable for fixed or mobile stations;
4) A laser-excited fluorescence system, using a Raman-shifted excimer laser and pulse compression techniques.

The air-borne UV LIDAR fluorosensor is an active remote sensing apparatus: using a laser source in visible-UV as transmitter, it allows to detect the amount and the type of pollutants and biological components dispersed in the sea, by means of the time- and wavelength-resolved measurement of the induced fluorescence light[21]. Moreover, in the frame of the ENEA participation to the national research program in Antarctica, the installation in the Italian base of a lidar station for monitoring ozone concentration is also foreseen.

II.3. Laser Sources

The area "Sources" has the objective to realize laser sources with features (wavelength, beam quality, mean and peak power) not available on the market, but important for application work, both in ENEA and in the industrial world. Activity lines have been defined in the field of:
- UV and VIS sources;
- IR sources;
- Free electron laser (FEL).

As to UV and visible sources, both high power and high beam quality laser sources are developed. Laser systems with high mean power in UV are used in the field of mechanical processing, especially of organic materials, where better results are obtained by UV sources than by IR sources, and for photochemical processes, where a high global efficiency is needed to make these processes competitive with respect to the traditional ones. The use of UV radiation, by itself or together with IR radiation, for mechanical working is very interesting both for the remarkable absorption coefficient and for the chance to concentrate high powers on very little surfaces, with a high precision. High peak power sources will be used in diagnostics of thermonuclear plasma, for microlithography and semiconductor annealing. The use of excimer lasers is the most suitable way to obtain laser sources tunable in visible. An excimer (XeCl) source, operating in UV (wavelength = 308 nm), with a high pulse energy (up to 11 J) has been projected, built and implemented[22]. Two sources will be implemented in the next future, with the following features:
a) repetition frequency 100 Hz, energy per impulse 1 J, mean power 100 W;
b) repetition frequency 10 Hz, energy per impulse 10 J, mean power 100 W.
This activity, included in the Eureka project Hipulse EU 213, has the final goal of realizing an excimer laser source of high mean power and high repetition rate (1 kW, 1 kHz) and a laser source with high energy per pulse (10 J/pulse, 100 pulses/sec).

In the IR region, CO_2 laser sources are being further improved. This type of laser has now a high reliability and can be used in many different applications, ranging from mechanical processing to medicine, photochemistry and remote sensing. Modifications and improvements are being studied in order to further extend its use: for example, it can be used to excite

52

other molecules and to obtain laser emission in other spectrum regions; this way, sources emitting in the microwave region have been realized. A pulsed CO_2 laser, with high peak power for the study of damage threshold of optical components has also been realized. Laser sources almost completely tunable in the far IR region (60-500 µm) by Raman effect and four waves mixing have been realized for photochemical applications[23]. Another goal is to develop tunable colour centre sources, in the wavelength region of 2 µm, with mean power of about 0.1 W, to be used for very high resolution spectroscopy, atmospheric pollutants trace measurements and study of transport properties in optical fibres[24]. A significant research program about colour centre lasers, with beam power modulation possibility for photochemistry, is operating, that includes:
- the improvement of colour centre lasers features;
- the realization and the engineering of prototypes;
- the study of fundamental optic properties of colour centres and impurities in insulating crystals;
- the production and the characterization of alkaline alogenides thin films (obtained by thermal vaporization) for a feasibility study of optoelectronic devices.

The Free-Electron Laser (FEL) is characterized by the novelty and the complexity of the elements needed for its realization, requiring the integration of capabilities developed in different fields, such as accelerators, magnet technology, etc. During the first 80's a large amount of theoretical and experimental work has been performed, relevant to the development of laser sources based on the stimulated synchrotron emission by high energy electron beams travelling in magnetic ondulators. The free-electron laser system has been subsequently completed and implemented in the ENEA Research Centre of Frascati at the end of 1985, while the relevant theoretical models and calculation codes are being developed[25]. This laser generates IR radiation at a wavelength of about 10 µm with a peak power of the magnitude of milliwatt. This laser does not produce a continuous radiation beam, but a series of light pulses of short duration (about $2 \cdot 10^{-11}$ sec) for an overall time of 5 µsec. The main feature of FEL, with respect to traditional laser sources, is that the active medium, instead of being made by atoms or molecules, is an electron beam of high energy, propagating through a magnetic structure (ondulator). The emitted radiation can be continuously tuned in a broad wavelength range. The final goal is to realize a tunable laser source in the 10-30 µm spectrum region, with a few percent efficiency, mean power of 50-100 watt, peak power in the Megawatt range and repetition rate of 100 pulses/second. The intensity and the duration of the pulse of the electron current will be raised by using a new source for the electron accelerator, while a new ondulator (made by permanent magnets) will be installed. In principle, the spectrum region can be widened (from visible to microwave) and both efficiency and power can be increased, by using electron beams of different energy and magnetic ondulators of new conception, at present under development. One of the most interesting features of FEL is the chance of continuously tuning the emitted radiation within a broad wavelength range (from 10 to 30 µm for the ENEA source), by varying some easily controlled parameters, such as the electron beam energy or the magnetic field of the ondulator. This fact makes this laser source very interesting in many application fields, among which photochemistry, surface studies, diagnostics and heating of nuclear plasma, telecommunications. Other FEL's are operating in USA and in France, or are under construction in Great Britain, Japan and Soviet Union, but the ENEA FEL is the only one operating with an accelerating machine of the microtron type. A Cerenkov-effect laser, continuously tunable from 100 um to 1 mm, where no tunable laser source is at present available, is also under development.

III. ENEA RESEARCH PROGRAM ON SURFACE TREATMENT

Other activities dealing with the applications of lasers in welding large thickness metals are being performed in the project "Surface Treatment of Metallic Materials". In this project, the deepening of knowledge about the beam-material interaction, the improvement of metal surfaces by laser and the relevant mathematical models are also studied. An "Application Center of Laser Beam" has been established by ENEA, together with three Italian companies, in order to provide assistance to national industry for the problems generated by the introduction of laser technology, particularly in welding. The problems faced by this Centre are: norms, material qualification, system realization, experimental testing, operating condition selection, staff training.

IV. ENEA RESEARCH ACTIVITIES IN THE FIELD OF ENVIRONMENT AND HUMAN HEALTH

Other ENEA activities involve the use of optoelectronic techniques: examples are the activities performed by the area Environment on remote sensing (in connection with the evaluation of the environmental impact of energy plants) and the use of lasers in flow cytometry. As to the first one, the thermal load of river Po has been studied, in the frame of an agreement signed by ENEA and the Lombardia Region, in order to study the impact of energy producing plants (particularly nuclear power plants) on environment and human health. The modern techniques of remote sensing in the thermal infrared are the most adequate means to collect the needed information about surface temperature of the river. An air-borne multi-spectral optomechanical scanner has been used, while the data have been evaluated by means of the ENEA Digital Imagery system (EDI)[26]. Flow cytometry is an analytical technique measuring features of cells as they flow past an observation point. In order to be characterized, particles must be illuminated with light of a sufficient intensity and appropriate wavelength, to detect scattering and fluorescence from natural and exogenous fluorochromes. Lasers are used because they provide high intensity, monochromatic, non-divergent beams. Flow cytometry is used at ENEA in the study of cytotoxic and cytogenetic effects of toxic agents[27,28], as well as in the characterization of heterogeneity of human solid tumour cells[29]. Moreover, the evaluation of the content of sub-cellular components in vegetable cells has been performed[30].

V. EUREKA PROJECTS

The growing diffusion of industrial applications of advanced technologies has led to several R&D initiatives at European level. ENEA coordinates the Italian participation in the Eureka initiatives.

V.1. Eurolaser

One of the most important among Eureka programs, Eurolaser, aims at the realization of high mean power lasers. The Eurolaser program includes eight projects: the following ones see a significant Italian participation, with a large ENEA contribution:
- EU 180 "CO_2 laser cells and related systems". The total cost of this project is 43 MECU, with an Italian share of 84%; the goal is to develop CO_2 laser sources with 10 kW modules and relative components in order to realize three laser station prototypes, for soldering, surface treatments and robotics respectively.
- EU 194 "Evaluation of industrial applications of high power lasers" . Total cost is 13 MECU, the Italian share 23%; the objective is to promote the applications of high power lasers (CO_2 lasers of power > 10 kW).
- EU 213 "High power excimer lasers". This project will cost 20 MECU, with an Italian share of 45%. The goal is to develop an excimer laser system of mean power more than 1 kW, repetition rate between 100 and 1000 Hz, and high flexibility for industrial applications.
- EU 204 "Laser workstation for surface treatments", (cost 18 MECU) aiming at developing a power laser system for materials surface treatments, with an optical multi-use automatic system and systems for controlling interfaces conditions with the production line.

Other Eureka projects, co-ordinated by ENEA but not including its direct participation are:
- EU 226 "High power solid state lasers" (cost 19 MECU) aiming at realizing a solid state laser system, with a power in the kW range, with good beam features and high efficiency, for industrial uses.

V.2. Euromar

The Eureka project EU 37 (Euromar), entitled "Development, application and successful exploitation of Europe's advanced marine technology having worldwide market potential", among other goals, aims at the development of remote sensing instrumentation and methods for use from satellite, aircraft and by radar techniques. Euromar is constituted by one pilot project (European Marine Surveillance and Information Network), including the core of problems dealt with by the whole project, and a number of subprojects. One of these sub-projects (Remote Sensing Systems) deals with the development of modular systems of remote sensors, both of passive (radiometers for visible, IR and microwaves) and active type (LIDAR, laser altimeters), for teledetection from aircraft, ship or land of oceanographic parameters that can be obtained by these techniques. A leading role in this project has been taken, at national level, by ENEA.

VI. ACKNOWLEDGEMENTS

The authors wish to acknowledge the assistance of many colleagues belonging to the ENEA Areas "Innovative Technologies" and "Environment" for the help in collecting all the cited information.

REFERENCES

1. U. Ascoli-Bartoli, S. Martellucci and E. Mazzucato, Nuovo Cimento, 32, 298 (1964)
2. S. Martellucci, Energia Nucleare, 18,10, 541 (1971)
3. A. Caruso, A. De Angelis, G. Gatti, R. Gratton and S.Martellucci, Phys. Letters, 29A,6, 316 (1969)
4. S. Scaglione and G. Emiliani, J. Vac. Sci. Technol. A3: 2702 (1985)
5. A. Piegari and E. Masetti, Thin Solid Films 124: 249 (1985)
6. G. Emiliani, E. Masetti and A. Piegari, SPIE 652:153 (1986)
7. E. Borsella, R. Fantoni and L. Caneve in: "Photon, Beam and Plasma Enhanced Processing" E-MRS Symp. Proc., Vol.XV, A. Galnsky, V.T. Nguyen and E.F. Krimmel eds., Les Edition de Physique, Paris 1987.
8. R. Fantoni, E. Borsella, S.Piccirillo and S. Enzo, J. Mat. Research 5, 143 (1990)
9. E. Borsella and R. Larciprete, Appl. Surface Sci. 36:221 (1989)
10. R. Larciprete and E. Borsella, Chem. Phys. Letters 147:161, (1988)
11. R. Fantoni, R. Larciprete, S. Piccirillo, G. Bertoni, R. Fratarcangeli and M. Rotatori, Chem. Phys. Letters, 143(3):245 (1988)
12. R. Engeln, R. Fantoni and G. Schina, Nuovo Cimento 12D, 209 (1990)
13. G. Baldacchini, A. Bizzarri, L. Nencini, M. Snels, J. Mol. Spectr. 130, 337 (1988)
14. G. Baldacchini, A. Bizzarri, L. Nencini, G. Buffa, O. Tarrini, J. Quant. Spectr. Radiat. Transfer. 42, 423 (1989)
15. A. De Angelis and F. Garosi, "Indagini strutturali sul cavallo di bronzo dei Musei Capitolini di Roma mediante interferometria olografica a doppia esposizione con laser impulsato di alta potenza", Proc. of 2nd Int. Conf. on "Non-destructive testing,microanalytical methods and environment evaluation for study and conservation of works of art", Perugia, April 1988, I/20 (in Italian)
16. A. De Angelis, P. Di Lazzaro, F. Garosi, G. Giordano and T. Letardi, Appl. Phys., B47, 1-6 (1988)
17. R. Barbini, F. Colao, A. Palucci and A. Petri, "The ENEA ground based CO LIDAR station",Proc. of the XIV International Laser Radar Conference, Innichen-S.Candido (Italy, June 20-24, 1988)
18. R. Barbini, A. Ghigo, M. Giorgi, K.N. Tyer, A. Palucci and S. Ribezzo, Optics Comm., 60:239 (1986)
19. R. Barbini, A. Ghigo, A. Palucci and S. Ribezzo, Optics Comm., 68:41 (1988)
20. E. Galletti, E. Stucchi, A. Ferrario, R. Barbini, P. Delli. G. Bitelli, F. D'Amato, M.Giorgi, "Developmemt of a CO pulsed laser for spaceborne coherent Doppler LIDAR", 5th Conference on Coherent Laser Radar, Munich, FRG, June 5-9, 1989, SPIE Proceedings, 1181:113 (1989)

21. R. Barbini, R. Fantoni, A. Palucci, S. Ribezzo and H.J.L. v.d. Steen, "A novel laser Fluorosensor: preliminary results", Proc. of the XV International Laser Radar Conference, Tomsk (USSR), July 23-27, 1990

22. T. Letardi, S. Bollanti, P. Di Lazzaro, F. Flora, G. Giordano, T. Hermsen and C.E. Zheng, "Study of 10 liter active volume, X-ray preionized XeCl discharge laser system", Proc. of the International Congress on Optical Science and Engineering, Hamburg, September 19-23, 1988

23. M. Bernardini, M. Giorgi, A. Palucci, S. Ribezzo and S. Marchetti, Europhys. Lett., 2:695 (1986)

24. G. Baldacchini, D. Censi, G.P. Gallerano, U.M. Grassano, M. Meucci, A. Scacco, M. Tonelli and P. Violino, Rev. Phys. Appl., 18:301 (1983)

25. U. Bizzarri, F. Ciocci, G. Dattoli, A. De Angelis, G.P. Gallerano, I. Giabbai, G. Giordano, T. Letardi, G. Messina, A. Mola, L. Picardi, A. Renieri, E. Sabia, A. Vignati, E. Fiorentino and A. Marino, Nucl. Instrum. & Meth., A250:254 (1986)

26. F. Borfecchia, P. Cagnetti, B. Della Rocca, "Thermal mapping of the Po-river by infrared remote sensing from aircraft: on field calibration", Proc. of the European Association of Remote Sensing Laboratories (EARSeL) Workshop, Capri, May 1988

27. M. Span, F. Pacchierotti, F. Mauro, S. Quaggia and R. Uccelli, Int. J. Radiat. Biol., 51:401 (1987)

28. M. Span, F. Pacchierotti, R. Uccelli, R. Amendola and C. Bartoleschi, J. Toxicol. Environ. Health, 26:361 (1989)

29. L. Teodori, D. Tirindelli Danesi, E. Cordelli, R. Uccelli, R. De Vita, M. Span, F. Mauro, A. Schillaci, A. Moraldi, L. Capurso and S. Stipa, "Potential Prognostic Significance of Cytometrically Determined DNA Abnormality in G.I. Tract Human Tumours", in: Clinical Cytometry, M. Andreeff, ed., N.Y. Academy of Sciences, New York (1986)

30. J. Dolezel, P. Binarova and S. Lucretti, Biologia Plantarum, 31:113 (1989)

ATMOSPHERIC OZONE

THE ITALIAN PROGRAM IN ANTARCTICA, RELATED TO THE OZONE HOLE

PROBLEM AND THE EXPERIMENTAL CLOUD LIDAR PILOT STUDY

L. Stefanutti, M. del Guasta, M. Morandi, V. M. Sacco,
F. Castagnoli, E. Palchetti, and L. Zuccagnoli
CNR-IROE, Firenze, Italy

G. Megie and S. Godin
Service d'Aeronomie du CNRS
Université Pierre et Marie Curie, Paris, France

Cai Peipei[+]
East China Normal University, Shanghai, China

I. INTRODUCTION

Since January 1988 an Italian backscattering Lidar has been operating
in Antarctica, at the Italian base of Terra Nova Bay during the summer
1988 and now at the French base of Dumont d'Urville. The system was built
by the research group of IROE of CNR, within a broad research program on
atmospheric studies carried out by the Italian National Program for
Antarctic Research. Among the principal objectives of this program are to
contribute understanding of the Ozone Hole problem and to characterize the
radiative properties of clouds by means of Lidar, IR radiometry, local
meteorological data and satellite data, in order to better understand the
role of clouds in the radiative budget of the planet. A complex DIAL
system, to carry out ozone concentration measurements in the Ozone Hole,
is presently being built under the Italian-French cooperative program for
Antarctica, while an automated Nd-YAG diode pumped Lidar is being designed
to operate semiautomatically in an Antarctic winter base, maybe at Dome C,
which will be located on the Antarctic Plateau, at 3250 meters altitude.

II. THE OZONE HOLE PROBLEM

In the cold Antarctic Stratospheric Vortex, when the temperature
drops below 195°K the nitric acid, present in the stratosphere as a
reservoir gas at altitudes between 15 and 25 km, freezes, forming the so
called Polar Stratospheric Clouds (PSC's) of Type I. Such clouds are
formed by small crystals, with typical dimensions of 1 μm. At temperatures
10 degrees lower, all the water vapour present in the atmosphere freezes,

---

[+]Visiting scientist at CNR-IROE with ICTP grant.

with the formation of PSC's of Type II. Such clouds are much more similar to cirrus clouds and are formed by large ice crystals, with sizes up to 100 μm, and their sedimentation time is of the order of the hours[1,2]. In such clouds (PSCs of Type I and II), heterogeneous chemical reactions occur, with consequent denitrification and a dehydration of the lower stratosphere. In such reactions $ClONO_2$, which is also a non-reactive reservoir gas produced by the reaction between ClO and NO_2, is destroyed by reactions occurring on the surface of the ice crystals of such clouds. The larger the crystal surface, the more efficient the heterogeneous reactions. Typical reactions are the following:

$$HCl(solid) + ClONO_2(gas) \dashrightarrow HNO_3(solid) + Cl(gas)$$

$$H_2O(solid) + ClONO_2(gas) \dashrightarrow HNO_3(solid) + ClOH(gas)$$

NO_x compounds are consequently removed from the gas phase and trapped in the PSCs. They may be then slowly removed from the stratosphere by sedimentation processes. Gaseous chlorine is at the same time liberated from reservoir gases, and upon the return of solar radiation with the advent of spring the Ozone Hole can occur[3].

Backscattering Lidar can characterize PSCs of Type I and Type II. From the Lidar signature their vertical distribution and optical properties may be evaluated; when using a linearly polarized laser, the depolarization ratio, permits identifying the two types of clouds. PSCs of Type I, formed by small particles (1 μm), are expected to introduce small depolarization, while PSCs of Type II (large particles up to 100 μm), should cause large depolarization ratios, as in the case of cirrus clouds.

II.1. Radiative Properties of Clouds - The ECLIPS Program

Tropospheric clouds have great importance in the radiation budget of the planet and hence affect climate, but their role is still not understood. A cloud may cause heating or cooling of the lower atmosphere, depending on its height, its optical thickness and radiative properties. Further, a modification of the cloud coverage may cause unknown climatic changes; man's modification of the environment might change the microphysical properties of the clouds, and consequently their radiative properties. An accurate knowledge of total radiative properties in the solar (0.3-40 μm) and terrestrial (4-50 μm) wavelength spectrum is of primary importance. The so called "infrared emissivity" ε of a cloud, defined as the ratio of the cloud's radiance to the radiance of a black body at the cloud's temperature, gives a good radiative description of clouds.

The ECLIPS (Experimental Cloud LIdar Pilot Study) program[4] is designed to provide a more complete description of clouds. Lidars can measure cloud base, cloud top, and optical depth. As in the case of PSCs, from the depolarization ratio the physical phase of clouds can be determined. Shortwave visible radiometers may give the total solar radiation between 0.3 - 2 μm which reaches the ground. Long wave IR radiometers and narrow-bandwidth, narrow field of view IR radiometers are suitable to measure the cloud emission in the IR region and hence the IR properties.

From all these data the cloud emissivity may be computed. Time lapse TV cameras may be used to derive cloud velocity and total coverage.

III. EXPERIMENTAL LIDAR PROGRAMS IN ANTARCTICA

Since December 1987 Italian Lidars have been operating in Antarctica. During the 3rd Italian Antarctic Summer Expedition the IROE backscattering Lidar was located in Tethys Bay at 74°41'42" S, 164°07'23" E. The Backscattering Lidar operated from December 30th, 1987 to February 10th, 1988 for a first series of tests, measuring stratospheric aerosols and tropospheric clouds[5,6]. Since January 1989 this system has beein operating on a routine basis at the French Base of Dumont d'Urville (66°40'S, 140°1'E) as a part of the joint Italian-French cooperation program in Antarctica.

A first mini ECLIPS study has been carried out during the month of April 1989. Polar Stratospheric Cloud measurements were the prime goal of the second Antarctic Lidar study.

III.1. The Elastic Backscattering Lidar

The polar elastic backscattering Lidar (Figure 1), designed at IROE, has the following main features[7]. The transmitter consists of a Quanta System Nd-YAG laser consisting of an oscillator, an amplifier and a second harmonic generator. The oscillator has a Self Filtering Unstable Resonator (SFUR), which produces an almost diffraction limited beam with 0.25 mrad semidivergence. The output energy is 0.4 J at 532 nm; the pulse repetition rate is 4 to 10 Hz. The receiver is formed by a double folded Newtonian telescope of 0.5 m diameter and a focal length of 2 m. A field diaphragm can be inserted in the focal plane providing variable fields of view from 0.25 to 0.6 mrad (semiaperture). After the collimator there is an interference filter with 0.15 nm bandwidth, followed by a cube polarizer to separate the two polarizations, and two photomultipliers followed by wideband video amplifiers.

Due to the very large dynamics of the Lidar signal, especially when measuring tropospheric clouds, it has been frequently necessary to vary the preamplifier gain and the output energy of the laser. To avoid, or at least reduce, this requirement new 100 MHz logarithmic video amplifiers by Optec will be introduced in 1990.

The generated electric signals enter two channels of a transient digitizer. Both 12 bit-5 MHz (LeCroy 6810) and 8 bit-32 MHz (LeCroy 8837) transient digitizers are available. The entire system is controlled by means of a personal computer (Olivetti M28 and M380). The photomultipliers have gate circuits acting on the first dynode, permitting us to reduce the PMT gain by a factor of about 800 for low altitude signals. The gating time can be varied in steps.

A second PC is available for data processing, as well as a vertically pointing wide angle TV camera for cloud monitoring, interfaced to a time lapse video recorder. For the second Lidar expedition a longwave Infrared

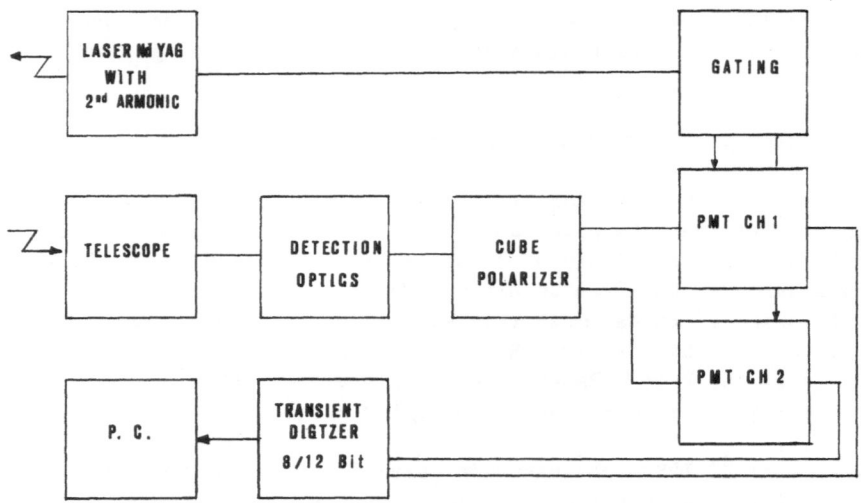

Fig. 1. Block diagram of the elastic backscattering Lidar system.

Radiometer (PIR) by Eppley was added to the system. For the third Lidar expedition a shortwave radiometer (PSP) will be added, and in the future an Infrared Radiometer PRT-5, with small field of view and limited band-width, especially calibrated for the cold Antarctic atmosphere.

The second PC must also perform the acquisition of the radiometric data and control the laser-receiving telescope alignment, by controlling the position of the output coupling mirror. The system is located inside an ISO 20" shelter, where particular attention was given to the thermal insulation. In the Antarctic environment the effect of the strong catabatic winds also has to be taken into account. Windspeeds of the order of 300 km/h have been measured at Dumont d'Urville. Under such circumstances the whole system goes into vibration, and alignment problems occur, both in the laser head and in the alignment between laser and telescope. For this reason the telescope and laser head are mounted on a mechanical structure which is insulated from the shelter's floor by means of vibroshock elements.

In order to make the lidar data available in Europe while the system is operating during the Antarctic winter, a computer satellite link has been established. A third PC has been connected by means of a modem to the INMARSAT ground station of Dumont d'Urville. A communication channel has thus been established with the Computer Center of the University of Singapore, where an account was opened. The data are transmitted at a rate of 1200 baud to Singapore and from there, via BITNET to the node at Florence. Finally, the data are transferred via modem to a PC at IROE, converted and sent, again via BITNET, to the Service d'Aeronomie of CNRS in Paris. Due to the high costs of the INMARSAT system, faster procedures and new modems are being tested, with the aim being to obtain transmission rates of 9600 baud and data compression. Most of the refurnishing of the system is being carried out with a view towards future automation, using simple procedures and automatic control. The lidar signature compression and research for the most reliable transmission channel are both fundamental elements for the eventual automated system.

IV. EXPERIMENTAL RESULTS

IV.1. Stratospheric Aerosols and Polar Stratospheric Clouds

The main goal of the 1987/88 Lidar Campaign in Antarctica was to test the instrumentation to be used in subsequent winter and spring operations. The first routine test was the one at Dumont d'Urville which started on January 8, 1989.

Two types of stratospheric lidar signatures have been obtained:
i) lidar profiles to monitor background stratospheric aerosols between 10 and 30 km;
ii) Polar Stratospheric Cloud profiles, measured in two polarizations at heights between 10 and 25 km during the Antarctic winter and early spring.

For all the stratospheric profiles the 12 bit-5 MHz transient digitizer has been used. The measurements of type i) above have been carried out only on the parallel polarization because of the presence of only a low background aerosol (probably H_2SO_4 with aerosol particle size below 1 μm), which does not introduce substantial depolarization. In the measurements for monitoring stratospheric aerosols the gating of the photomultiplier was regulated up to 7 km. The measurements were averaged over 5000 shots during the summer time (day time measurements), and reduced to 500 shots when night measurements could be performed. These signatures have been normalized versus an average molecular atmosphere derived by the daily radiosonde data produced by the Meteorological Office of the Base. Small aerosol layers have been detected between 10 and 25 km. The integrated backscattering and the scattering ratio R, defined as

$$R = \frac{\beta_{Mie} + \beta_{Rayleigh}}{\beta_{Rayleigh}} = 1 + \frac{\beta_{Mie}}{\beta_{Rayleigh}}$$

have been computed.

Also comparison between the lidar data and the SAM 2 satellite data have been performed. To this end the following assumptions were made: a $1/\lambda$ dependency for the aerosol Mie scattering and extinction to a backscattering ratio of 40. Such a value for ∂/β was determined by imposing a coincidence of the Lidar and the SAM 2 data at 18 km for one single measurement. Consequently, a very good coincidence was obtained for the whole signature for all measurements examined. The comparison was performed by calculating the lidar extinction values at 1 μm (the wavelength at which SAM 2 operates).

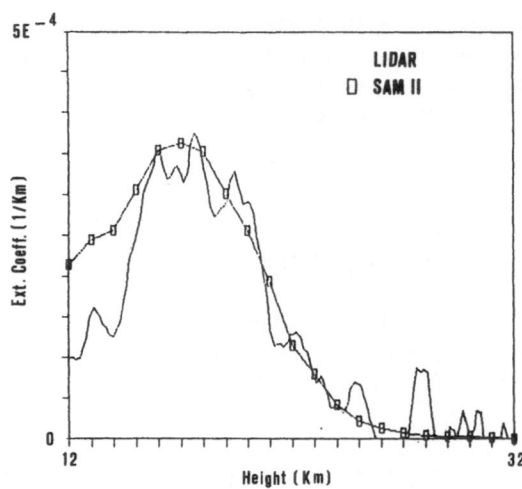

Fig. 2. Lidar versus SAM II comparison:
Aerosol measurements of February 2,
1989. Comparison of the extiction
coefficient due to Mie scattering
at 1 μm as obtained by Lidar and
by SAM 2.

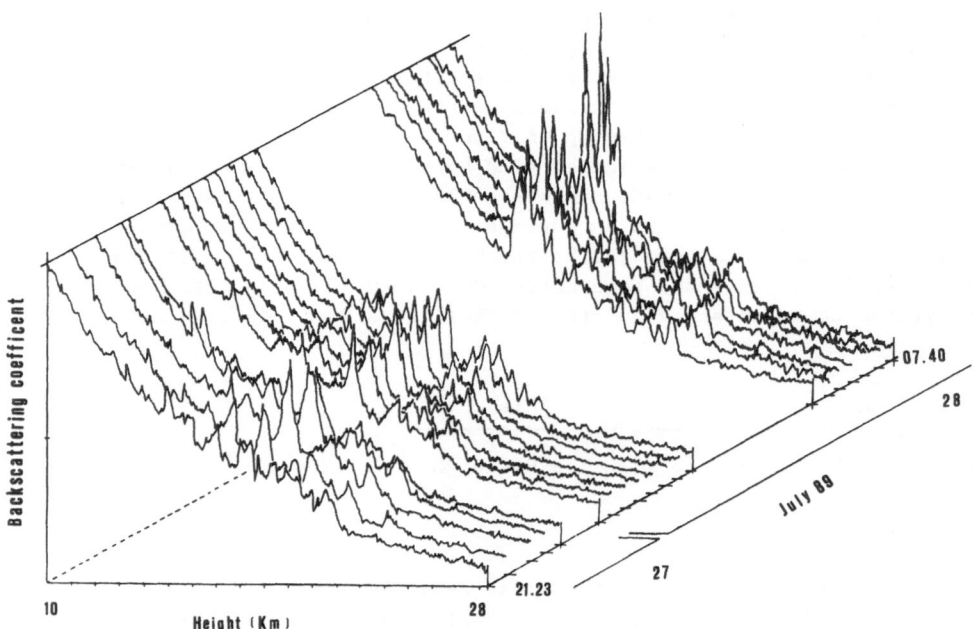

Fig. 3. Time evolution of Polar Stratospheric Clouds; in abscissa
there is the altitude, on the y axis the backscattering
coefficient and on the z axis the time. The figure
represents a 10 hours evolution. Different layers of PSCs
may be observed. Sedimentation processes of the order of
100-150 m/h can be evaluated.

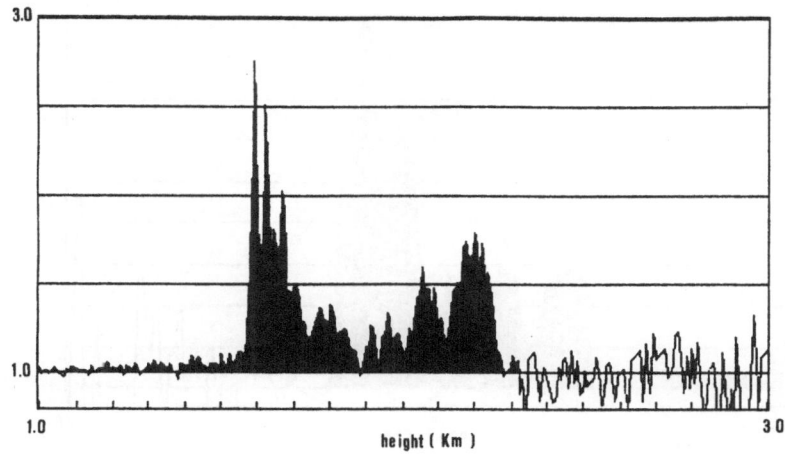

Fig. 4. Scattering ratio - pol p (Dumont d'Urville, 66°40'S,
140°1'E, July 28, 1989, time 7h40', max scatt. ratio =
= 2.76015 at 15.9 Km). The scattering ratio versus height
is plotted for one of the measurements of Figure 3 for
the polarization parallel to the one of the emitted
laser pulse. Values of R of the order of 3 could be
evidenced.

Figure 2 shows one of such comparisons. The vertical resolution of
the Lidar is much better than that of the satellite. The maximum scattering ratio measured over the whole period, when PSCs were not present,
never exceeded the value of 0.1. Most likely, the measured aerosol is
formed by a solution in water of sulfuric acid generated by volcanic
eruptions or by injections into the stratosphere of biogenic materials
such as OCS, derived from the destruction of organic materials in the
sea[8].

These measurements constitute a baseline for further research. During
the Antarctic winter of 1989 there has been a continuous monitoring of PSCs.
To this end measurements on both polarizations have been performed.

Polar Stratospheric Clouds were detected only very late at Dumont
d'Urville; heavy loading of the stratosphere by PSCs was detected starting
in the second half of July. Figure 3 shows a three-dimensional description
of such a phenomenon. From the figure it appears that the loading evolves
quite rapidly in time. Such measurement, obtained by conducting groups of
500 shots at 4 Hz every ten minutes, permit us to evaluate possible sedimentation processes. From an analysis of Figure 3 one may notice that the
layers present between 15 and 23 km tend to lower with time; this is
probably due to a sedimentation process. At a first evaluation, for the
case under examination one may deduce a sedimentation velocity of about
0.1 km/h.

Figure 4 shows the scattering ratio obtained for one single group of
measurements. The scattering ratio R is much stronger than in the case of
background aerosol, and very similar to the case of strong volcanic aerosol
loading of the stratosphere. Values of R larger than 3 have been measured.

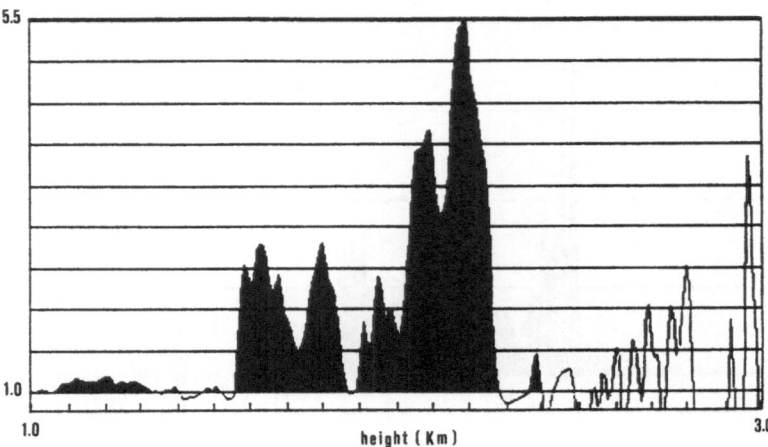

Fig. 5. Scattering ratio - pol. s (Dumont d'Urville, 66°40'S
140°1'E, July 28, 1989, time 7h 40', max scatt. ratio =
= 5.500731 at 21.9 km). The scattering ratio versus height
for the depolarization channel. The difference from Figure
4 may be noticed. From the data of Figures 4 and 5 the
depolarization ratio can be computed.

Figure 5 shows the same lidar profile measured on the crossed polarized component. The change in the scattering ratio R is evident. This is
due to the depolarization caused by the crystal particles which form the
PSCs. Depolarization values up to 25% were found in PSCs. Such high values
probably indicate large particles and therefore PSCs of Type II. Values of
the depolarization of the order of 5-10 % were also measured. In this case
it is more probable that there is the presence of PSCs of Type I.

The measurements of PSCs of Type II might be a consequence of transport from inside the continent as, from the radiosonde data available,
stratospheric temperature as low as 283 °K were never reached. The radiosonde data are of extreme interest in the study of such phenomena, since
from the temperature of the stratosphere it is possible to deduce, for a
location such as Dumont d'Urville at the edge of the Antarctic continent,
whether the station is inside or outside the Polar Vortex.

For a more detailed study of the events data on the Polar Stratospheric Circulation should be also used. Furthermore, with the intensity
of the lidar signature of PSCs it becomes reasonable to invert the
signature to compute the extinction and the backscattering coefficient in
the cloud and obtain the optical parameters of PSCs. The method will be
illustrated in the next paragraph, in the discussion of tropospheric cloud
measurements.

IV.2. Tropospheric Clouds - The ECLIPS Experiment

For stratospheric clouds and also for Cirrus clouds it is generally
possible to assume that outside the cloud there is only a molecular atmosphere. Therefore, if the laser pulse penetrates all through the cloud,

and this always happens in the case of cirrus clouds for a powerful lidar system[7], it is possible to normalize the lidar signature below and above the cloud with an antarctic molecular atmosphere, which can be determined directly by radiosonde data. In this manner, we can compute directly the transmission change due to the cloud.

The signal before the cloud (at cloud bottom) will be:

$$P(R) \cdot R^2/k = \beta(R) \exp \int_0^R \partial_{Rayleigh}(r)\, dr$$

and the signal after the cloud (at cloud top) will be:

$$P(R+\Delta R) \cdot (R+\Delta R)/k = \beta(R+\Delta R) \exp \int_0^{R+\Delta R} \left[\partial_{Rayleigh}(r) + \partial_{Mie}(r) \right] dr.$$

Either the Klett method or a simple step by step computation can be used. This latter, developed by our group, permits us to directly compute ∂ and β for homogeneous clouds. The value of ∂ and β for the molecular atmosphere at the cloud bottom is known. Then by means of a step by step computation with few iterations one can compute both the extinction coefficient and the backscattering coefficient in the cloud assuming a linear relation between backscattering and extinction coefficients[8].

The method consists in assigning an arbitrary value A to the ∂/β ratio, by integrating the lidar signal step by step inside the cloud up to cloud top. If the assigned value A is correct, at the top of the cloud one will obtain the same value derived by normalizing the signal by means of the molecular atmosphere just after the cloud. If this is not the case, the value of A might be decreased or increased until the right value is obtained. This method seems to give good results as long as multiple scattering can be neglected and when there is still sufficient signal after the cloud. The single scattering assumption is reasonable when clouds are optically not very thick and laser transmitters and receivers have very small receiving field of view (smaller than 1 mrad full aperture).

Extensive statistical data on tropospheric clouds have been collected since January 1, 1988 at the Italian Base of Terra Nova Bay. The Lidar system has operated over the 24 hours in a burst mode: groups of 50 shots at 4 Hz were averaged and the time spacing between successive groups was one minute. This long sequence of measurements was interrupted only in the presence of perfectly clear sky, when stratospheric measurements were performed. The Lidar was operated in a two channel mode in order to measure also the crossed polarized components of the lidar signature. A vertically pointing TV camera recorded the evolution of the cloud patterns.

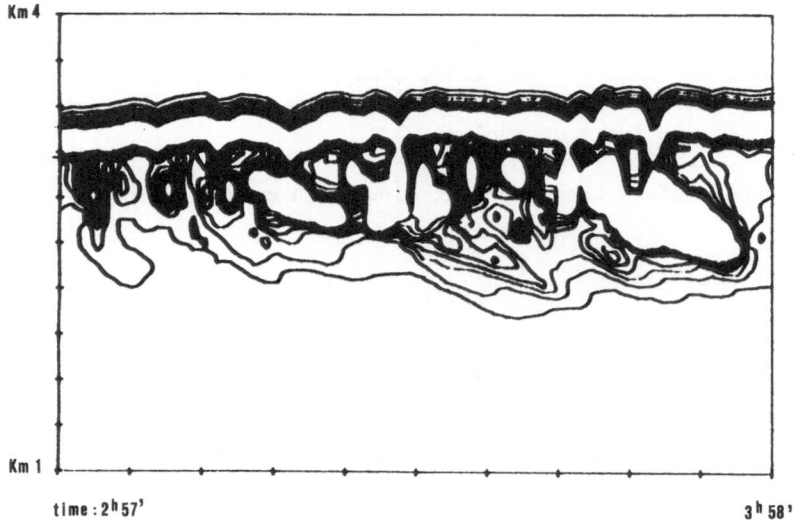

Fig. 6. Terra Nova Bay (February 2, 1988). The mixing ratio
(R-1) for a tropospheric cloud. The figure represents
one hour evolution.

By means of both camera pictures and cloud base height lidar measurements it was possible to compute the cloud velocity vector, and an evaluation of the cloud coverage was possible.

Most of the time low level Stratus clouds were observed. In most cases these Stratus clouds seemed to be optically thin, as higher level clouds could be monitored at various altitudes. Extinction coefficient, backscattering coefficient, optical thickness, cloud base and cloud top have been computed according to the method indicated above. Isoplets curves have been performed both for the mixing ratio (R-1) and for the depolarization ratio; examples are shown in Figure 6 and 7.

From a first partial analysis of the data the following results have been obtained: both water clouds and ice clouds have been monitored in Antarctica. Generally the low level Stratus were formed by water (depolarization ratio never above 10%), while medium clouds and high clouds always presented very high depolarization ratioes, of the order of 30 to 50%. In the case of low layers the presence of strong depolarization ratios has been often observed below the cloud base. Furthermore, from the time evolution of the phenomena a possible sedimentation time for ice particles could be measured (Figure 7), leading to crystal dimensions[9] of the order of 1 mm.

Many such techniques developed for tropospheric clouds have been used[10] in the characterization of the PSCs.

The Lidar measurements of troposheric clouds are presently being carried out within the Experimental Cloud LIdar Pilot Study (ECLIPS). In April 1989 a minicampaign carried out by NASA in Bermuda and our station in Dumont d'Urville took place. Routine measurements with one lidar shot

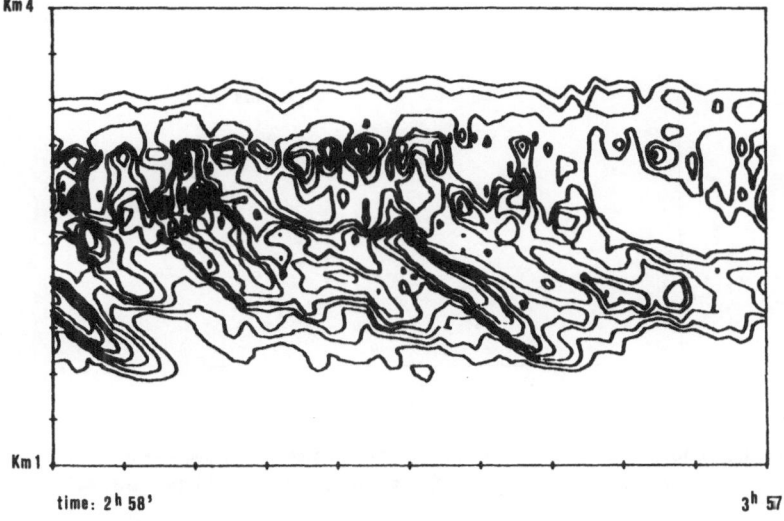

Km 4

Km 1

time: 2ʰ 58'   3ʰ 57'

Fig. 7. Terra Nova Bay (February 2, 1988). The depolarization
ratio for the same case of Figure 6. Sedimentation
processes of ice crystals may be noticed under the
cloud bottom. Crystals of the sizes of 1 mm could be
evidenced.

every minute are carried out all year around. The aim of the ECLIPS
program is the study of the effects of clouds on the radiation budget. The
Lidar system is now equipped with an Eppley IR longwave radiometer (PIR),
to measure the infrared emission of the clouds in the region between 5 and
50 μm. Because the measurements are carried out on the ground, the detected
radiance value is the sum of the cloud and atmospheric radiance:

$$L_M = \tau \cdot (I_C + I_G) - L_S$$

where M, C, S and G indicate respectively: measured, cloud, sky and
ground, and τ is the atmospheric transmission.

It is necessary to eliminate the contribution of atmospheric radiation and absorption due to water vapour and carbon dioxide between the
cloud and earth. The LOWTRAN 6 computer code[11] has been used to calculate
the theoretical downward radiance including clouds and atmospheric constituents. From the computed radiance and the measured values the infrared
emissivity ε of the cloud can be obtained. Lidar provides the cloud geometric parameters and the radiosonde data give the temperature and humidity of every atmospheric layer. Since the IR radiometer has a 170° field
of view, hemispherically incident radiance must be considered. Slant path
radiation and absorption with different zenith angle were computed and a
weighted sum was performed. Integrated radiance was needed due to the
wide band characteristics of the IR radiometer.

For obtaining more precise the IR emissivity of the cloud, the cloud
backscattering coefficients provided by lidar should be introduced. Neglecting cloud scattering in the infrared region, the IR emissivity ε is

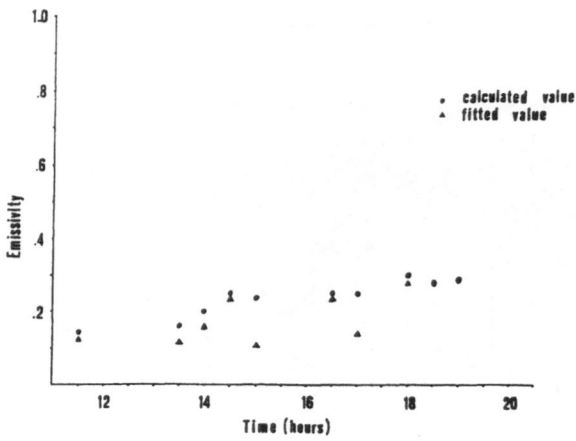

Fig. 8. The Cloud Emissivity versus time as computed
from the long wave IR radiometer and lidar data.

related to the infrared absorption depth as follows[12]:

$$\varepsilon = 1-\exp[-\delta_A(h)] \qquad (1)$$

if $\delta_C(h)= \alpha\delta_C(h)$, where $\delta_C(h)$ and $\delta_A(h)$ are respectively the visible
and infrared optical depth and α is a constant. Eq. (1) can be written[12]

$$J'(\Pi) = (k/2) [1 - \exp(-2 \alpha\ln 1/(1-\varepsilon))] \qquad (2)$$

where $J'(\Pi)$ is called the integrated backscattering and can be calculated
from the lidar data. Both k and α can be obtained through a least squares
fitting technique of Eq. (2) using lidar data and the value ε obtained by
means of the LOWTRAN 6 computation. Finally, a more accurate value of ε can
be calculated from Eq. (1).

Figure 8 shows the IR emissivity ε of a cloud as a function of
time, measured at Dumont d'Urville on February 2nd 1989.

A better calibration technique of the Eppley radiometer is planned
for 1990 and a radiometer operating in the visible will implement the
Lidar station. In 1991 also a PRT-5 narrow band, (spectral range 8-14 μm),
and narrow field of view (2° aperture) radiometer will be installed at
Dumont d'Urville.

V. THE AUTOMATIC BACKSCATTERING LIDAR SYSTEM

The principal disadvantage of the present backscattering system is
the necessary presence of an operator who has to control and decide the
type of measurement to be performed: stratospheric or tropospheric cloud
measurements. In the case of a clear atmosphere stratospheric measurements
have to be performed, and in the case of cloud coverage tropospheric cloud
measurements. The use of the IR longwave and visible radiometers connected

to a personal computer permits one to distinguish between clear sky and cloud coverage and between day and night. Hence it is possible to write a computer program which automatically selects the type of measurement as a function of the radiometers' output. There is little problem in setting the lidar to operate for the measurement of the stratosphere; once the dynamic range of the digitizer is set, this does not change with time. Only the number of shots to average may be changed as a function of the elevation of the sun over the horizon, which can be computed automatically.

Tropospheric cloud measurements are the main problem, due to the sharp changes of the dynamic range of the signal. To this end on the present system logarithmic amplifiers are being experimented to compress the lidar signature. A computer program can easily be written with a number of options for the sequence of different types of measurements as a function of time. The main difficulty today consists in the unreliability of the lasers and their high power consumption. At full energy and at a pulse repetition rate of 10 Hz typical consumptions of 4-5 kW have to be taken into account for the laser transmitter alone. The use of flashlamp pumping techniques for the Nd-YAG laser leads also to a limited lifetime, in the range of 10 million shots. For an automatic station a new philosophy must to be adopted. Nd-YAG diode pumped lasers will be tested, developing techniques presently proposed for space missions.

V.1. The Diode Pumped Nd-YAG Laser

A research contract has just being submitted to Quanta System for the development of a diode array pumped Nd-YAG pulsed laser. One dimensional arrays of diodes will pump the active crystal. The system will be formed by an oscillator and one amplifier with the following features:

Pulse repetition rate   100 Hz

Output energy at 532 nm   10 mJ

Second Harmonic Generator   BBO with 50% efficiency

Wall plug efficiency   > 7%

The oscillator will be pumped by one or two diode arrays, and the amplifier by 3 or 4 arrays. New active materials will be tested, as for instance Nd:YLF and LNA, to evaluate the most suitable material. The expected optical efficiency should reach values above 20%. The laser should be able to operate continuously for over a year, with a total power consumption of the order of 50 W. With such a solution the energy consumption is no longer a problem and automatic operation becomes feasible.

VI. THE DIFFERENTIAL ABSORPTION SYSTEM

Measurements of the ozone concentration by means of the DIAL system are possible both in the Ozone Hole and in the upper stratosphere. The main problem in carrying out precise ozone concentration measurements by means of the DIAL technique in the Antarctic Ozone Hole relies on the fact that there are two extremely perverse factors to be considered: the very low ozone concentration[13], and the high concentration of PSCs at elevations between 15 and 25 km. Classical DIAL techniques to measure ozone by means of two wavelengths, one inside the ozone absorption band and one outside, typically at 308 and 355 nm, but these lines are spaced too far apart. The presence of PSCs may affect the two lines differently, which is not easy to model as the scattering function of the aerosol particles is not known "a priori". Consequently no reliable O_3 concentration measurement may be performed. Moreover, two lines more closely spaced inside the ozone absorption band, as proposed by Pelon and Megie[14], typically 289 and 299 nm, might be not sufficient. It might be necessary to use at least 3 wavelengths to take into account the effect of the aerosols. Only for the upper Stratosphere, above 30 km, in the absence of aerosols, is it possible to measure Ozone with a traditional DIAL technique.

A complex DIAL is presently being assembled at IROE. Several studies on Ozone DIAL systems have been carried out during the past years in order to define the optimal configuration for the measurement of low ozone concentrations in the presence of aerosols. Such studies have been developed in Europe also within the EUROTRAC program. Such studies are of great help for the design of an Antarctic system. The DIAL developed at IROE in cooperation with the Service d'Aeronomie of CNRS is a double system, formed by a so called Tropospheric Ozone Lidar and a Stratospheric Ozone Lidar, the first to operate up to 20 km, the second from 20 to 50 km. The DIAL system will use a Quanta System Nd-YAG laser operating in the fourth harmonic, which will pump two Raman shifters, filled respectively with H_2 and D_2, to obtain the 289 nm and the 299 nm lines. Also the second harmonic radiation at 532 nm is being emitted. In the "tropospheric mode" the 289 and 299 nm radiation will be collected. For an evaluation of the aerosol distribution and PSC's distribution, measurements with the 532 nm line will also be carried out, subsequent to the UV measurements. By inverting the 532 nm data, which will be collected in two polarizations, important information also on the size distributions and concentrations of the aerosols may be derived, as is necessary to perform the evaluation of the ozone concentration.

For upper atmosphere measurements an EMG 150 MSC Lambda Physik Excimer laser is also being used. Such a laser, in the injection locked configuration, will produce only the 308 nm pulse, at high repetition rate (50 Hz) and low divergence. The 355 nm will be produced by the Nd-YAG laser, by frequency mixing of the fundamental and the second harmonic. In such a mode a prism is inserted into the harmonic generation system which inhibit the generation of the fourth harmonic. The receiver is a wide .8 m in diameter Cassegrain aluminum telescope. The visible radiation is coupled on a separate channel by means of a dichroic mirror, while all the UV radiation is fed to a high resolution spectrometer with a grating with 3600 r/mm, which is the present technology limit.

All alignment procedures are computer controlled. Visible and longwave IR radiometers will be mounted on the roof of the system. Also the

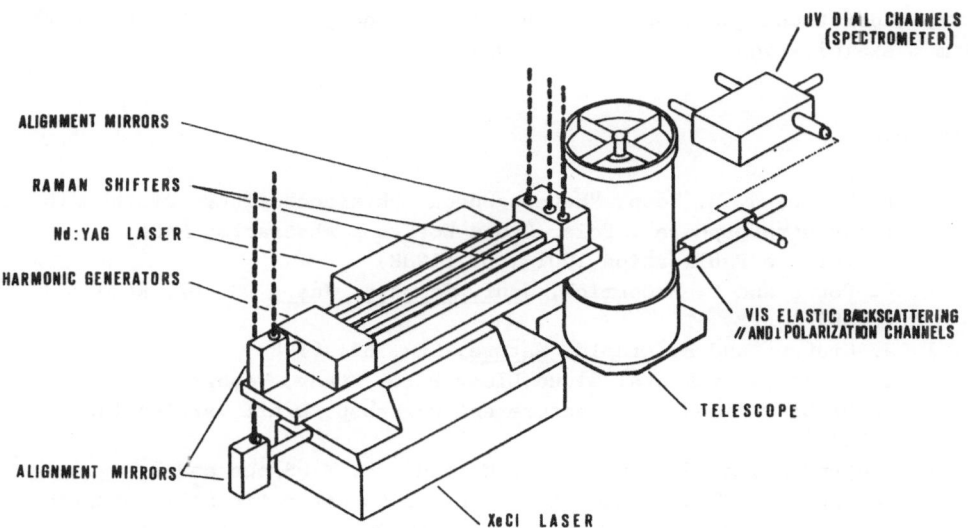

Fig. 9. Schematics of the DIAL system.

narrow field of view, narrow band IR radiometer, PRT-5, will be hosted in this system to measure cloud temperature. Figure 9 shows a schematic of the system which is hosted in two jointed ISO 20 aluminum shelters. A large quartz window will separate the telescope from the outside atmosphere.

The DIAL system will perform the following measurements:
i)   Ozone concentration up to 20 km in the tropospheric mode, inside the Ozone Hole;
ii) Ozone concentration in the high Stratosphere and lower Mesosphere, to determine if there is evidence of homogeneous chemistry of ozone depletion in the higher atmosphere;
iii) PSC measurements and their characterization by means of the depolarization technique;
iv) Determination of the PSCs temperature by means of the PRT-5 radiometer; and,
v)   Determination of the temperature of the upper Stratosphere and Mesosphere by means of the inversion of the lidar signal of the molecular atmosphere. Such measurements will be carried out by means of the 355 nm radiation [15].

Such a system will be tested during 1989 and 1990. It will become operational at Dumont d'Urville beginning from January 1991. At present the DIAL is scheduled to operate for at least 5 years in order to be used in climatic change studies.

VII. CONCLUSIONS

As a final remark we want to underline the necessity of international cooperation in such large programs and the opportunity to unite both mature lidar techniques and new sophisticated ones, in order to obtain the maximum exploitation of the sources available.

REFERENCES

1. R. P. Turco, O. B. Toon, "Heterogeneous Physicochemistry of the winter polar stratosphere", Polar Ozone Workshop Abstracts, NASA Conference Publication 10014, 61 (1988)
2. L. R. Poole and P. McCormick, Jour. Geophys. Res., 93, D7, 8423 (1988)
3. P. J. Crutzen and F. Arnold, Nature, 324, 651 (1986)
4. ECLIPS - An Experimental Cloud LIdar Pilot Study, Report of the WCRP/CSIRO cloud base measurement workshop, Draft version (June 1988)
5. L. Stefanutti, M. del Guasta, M. Morandi, V. M. Sacco and L. Zuccagnoli, "The application of Lidar to Antarctic Stratospheric Research", Proc. 1st Workshop Italian Research on the Antarctic Atmosphere, M. Colacino, G. Giovanelli and L. Stefanutti, eds., Porano, November 11th 1988, p. 161, S.I.F. Editrice Compositori, Bologna (1989)
6. M. Morandi, L. Stefanutti, M. del Guasta, E. Palchetti and V. Venturi, "Antarctic Tropospheric Cloud Characterization using an Elastic Backscatter Lidar", Proc. 1st Workshop Italian Research on the Antarctic Atmosphere, M. Colacino, G. Giovanelli and L. Stefanutti, eds., Porano, November 11th 1988, p. 161, S.I.F. Editrice Compositori, Bologna (1989)
7. V. M. Sacco, F. Castagnoli, M. Morandi and L. Stefanutti, Optics and Quant. Electron., 21, 215-226 (1989)
8. R. P. Turco, R. C. Whitten, O. B. Toon, J. B. Pollack and P. Hamill, Nature, 283 (1980)
9. R. H. Dubinski, A. I. Carswell and S. R. Pall, Appl. Opt., 24, 1614 (1985)
10. A. Heymsfield, Jour. of. Atm. Sci., 29, 1348 (1972)
11. F. X. Kneizys, E. P. Shettle, W. O. Gallery, J. M. Chetwind, L. W. Abrev, J. E. A. Selby, S. A. Clough and R. W. Fenn, "Atmospheric Transmittance/Radiance: Computer Code LOWTRAN 6", A FGL-TR. 83-0187, Environmental Research Paper N. 846, Air Force Geophys. Lab. (1983)
12. C. M. R. Platt, Jour. Appl. Meteor., 18, 1130 (1979)
13. D. J. Hofmann, J. W. Harder, J. M. Rosen, J. Hereford and J. Carpenter, "Ozone Profile Measurements at McMurdo Station Antarctica during the spring of 1987", Proceedings of the Polar Ozone Workshop, NASA Conference Publication 100014 (May 1988)
14. J. Pelon and G. Megie, Journ. Geophys. Res., 87, C7, 4947 (1982)
15. A. Hochercorne and M. L. Chanin, Geoph. Res. Lett., 7, 8 (1980)

DIFFERENTIAL ABSORPTION LIDAR DETECTION OF OZONE

IN THE TROPOSPHERE AND LOWER STRATOSPHERE

E. V. Browell

Atmospheric Sciences Division
NASA Langley Research Center
Hampton, Virginia, U.S.A.

I. INTRODUCTION

The distribution of ozone (O_3), in the atmosphere has important implications for pollution, chemistry, and climate related processes in the troposphere and for the screening of solar ultraviolet radiation, chemistry, and thermal equilibrium in the stratosphere. The large spatial variability of O_3 in the troposphere and in the vicinity of O_3 depletion episodes in the stratosphere makes in situ sampling of O_3 from balloons and aircraft inadequate for these types of O_3 investigations. The Differential Absorption Lidar (DIAL) technique has been used since the early 1970's for ground-based remote measurements of O_3 and aerosols in the lower atmosphere. This has permitted frequent measurements of O_3 and aerosols in the lower atmosphere. This has permitted frequent measurements of O_3 profiles above or around a specific location, which is an improvement over infrequent balloon measurements and even more infrequent aircraft measurements. To address the regional- and large-scale processes associated with O_3 transport and chemistry, an airborne DIAL system was developed in 1980 at the NASA Langley Research Center (LaRC) for the measurement of O_3 and aerosol distributions along the aircraft ground track. This DIAL system currently has the capability to measure O_3 and multiple-wavelength aerosol profiles to a range of over 8 km above and below the aircraft simultaneously. Over 12 major field experiments have been conducted with the NASA airborne DIAL system since 1980 to study the transport and chemistry related to O_3 and aerosols.

This paper discusses the DIAL technique for deriving O_3 profiles from lidar measurements. The NASA airborne DIAL system is described as an example of an advanced field system, and results are presented for studies of: a) photochemically produced O_3 in the summertime over the eastern United States and in biomass burning plumes during the dry season over the Amazon Basin of Brazil; b) vertical O_3 transport from the mixed layer into the free troposphere via cloud dynamics and from the stratosphere into the troposphere via tropopause fold events; and, c) O_3 depletion in the O_3 hole over

Antarctica. The airborne DIAL measurements discussed in this paper demonstrate the advanced capability of lidar for conducting O_3 investigations throughout the troposphere and lower stratosphere under widely different atmospheric conditions.

II. DIFFERENTIAL ABSORPTION LIDAR TECHNIQUES

The Differential Absorption Lidar (DIAL) technique for the remote measurements of gas profiles has been discussed by numerous authors[1-6]. In the DIAL technique, the average gas concentration over some selected range interval is determined by anlyzing the lidar backscatter signals for laser wavelengths tuned "on" (λ_{on}) and "off" (λ_{off}) a molecular absorption peak of the gas under investigation. A simplified version of the DIAL concept is shown in Figure 1. The value of the average gas concentration N_A (cm^{-3}) in the range interval from R_1 to R_2 can be determined from the ratio of the lidar signals at the on and off wavelengths, where $\sigma_A(\lambda_{on})-\sigma_A(\lambda_{off})$ is the difference between the absorption cross sections at the on and off wavelengths, and $P_{ron}(R_i)$ and $P_{roff}(R_i)$ are the signal powers received from range R_i at the on and off wavelengths, respectively. The average gas mixing ratio in parts per billion by volume (ppbv) can then be determined by dividing N_A by the local molecular number density for air. It is assumed in this analysis that the optical properties of atmospheric aerosols and molecules are equal at the on and off DIAL wavelengths. If

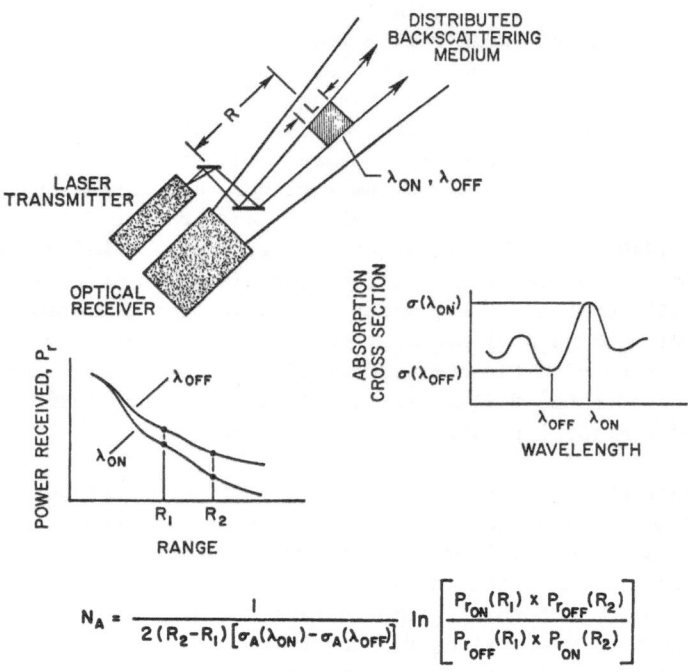

$$N_A = \frac{1}{2(R_2-R_1)\left[\sigma_A(\lambda_{ON})-\sigma_A(\lambda_{OFF})\right]} \ln\left[\frac{P_{r_{ON}}(R_1) \times P_{r_{OFF}}(R_2)}{P_{r_{OFF}}(R_1) \times P_{r_{ON}}(R_2)}\right]$$

Fig. 1. Differential Absorption Lidar (DIAL) concept.

there is an interfering gas which does not have the same absorption coefficient at these wavelengths, the concentration of this gas must be known or be determined by a separate measurement. In addition, a significant adjustment must be made to the calculated gas concentration when the wavelength dependence for total backscattering is a strong function of signal range. In particular, most simulation studies make use of the fact that these interference errors approach zero as the difference in DIAL wavelengths becomes small[7-9].

A general form of the DIAL equation can be derived which does not make any assumptions about the aerosol and molecular wavelength dependence of backscattering and extinction[5]:

$$N_A = \frac{1}{2(R_2 - R_1)\Delta\sigma} \ln \frac{P_{r\ on}(R_1)P_{r\ off}(R_2)}{P_{r\ off}(R_1)P_{r\ on}(R_2)} \qquad (M)$$

$$- \frac{1}{2(R_2 - R_1)\Delta\sigma} \ln \frac{\beta_{on}(R_1)\beta_{off}(R_2)}{\beta_{off}(R_1)\beta_{on}(R_2)} \qquad (B)$$

$$- \frac{1}{\Delta\sigma} (\alpha_{on} - \alpha_{off}) \qquad (E) \qquad (1)$$

where $\Delta\sigma = \sigma_{on} - \sigma_{off}$, β is the volume backscattering coefficient, and α is the average extinction coefficient between R_1 and R_2. The (B) and (E) terms in Eq. (1) amount to interferences in dthe DIAL measurement of N_A from wavelength variations in backscattering and extinction, respectively. The (M) term in Eq. (1) gives the traditional DIAL measurement provided that the error terms (B) and (E) can be neglected. The error term (B) can be neglected under conditions of spatially homogeneous backscatter, and the magnitude of term (E) is typically smaller than (B): however, it cannot be neglected under conditions of low atmospheric visibility.

It is usually assumed that DIAL extinction errors resulting from molecular scattering and absorption by interfering gas species can be reduced through simple approximations of the atmospheric gas density profiles and appropriate choices of the DIAL on and off wavelengths. The remaining extinction error is then due to a finite wavelength dependence of the optical attenuation by atmospheric aerosols. Given that the wavelength dependence of the aerosol extinction coefficient can be approximated by the power law $\alpha_{aer,\lambda} \cong \lambda^{-\gamma}$ where γ is the so-called Angstrom coefficient, then the extinction error is

$$E \cong \frac{-\gamma\ \alpha_{aer,\lambda_{on}}}{\Delta\sigma} \frac{\Delta\lambda}{\lambda_{on}} \qquad (2)$$

where $\Delta\lambda = \lambda_{off} - \lambda_{on}$. The Angstrom coefficient coefficient γ for atmospheric aerosols at visible wavelengths under conditions of low relative humidity is on the order of unity[10]. At high relative humidities the Angstrom coefficient approaches zero for most aerosol models, and extinction errors in DIAL measurements of O_3 can be neglected. Clearly, the extinction error due to the wavelength dependence of the aerosol attenuation will be reduced as the wavelength separation of the on and off wavelengths is reduced.

The backscatter error for DIAL measurements can be rewritten in terms of the so-called backscatter mixing ratio[11], $S\lambda(R)$, such that

$$B = \frac{1}{2(R_2 - R_1)\Delta\sigma} \ln \frac{|1 + S_{on}(R_1)|}{|1 + S_{off}(R_1)|} \frac{|1 + S_{off}(R_2)|}{|1 + S_{on}(R_2)|} \qquad (3)$$

and the backscatter mixing ratio is defined by

$$S(R) = \frac{|P_{aer,\pi}(R)/4\pi|\sigma_{aer}(R)}{(3/8\pi)\sigma_{mol}(R)} \cong S_{\lambda_0}(R) (\frac{\lambda}{\lambda_0})^{4-\delta} \qquad (4)$$

where:

$P_{aer,\pi}(R)/4\pi$ — normalized aerosol backscatter phase function ($\propto \lambda^{-\varepsilon}$ where ε is the wavelength dependence parameter)

$\sigma_{aer}(R)$ — aerosol scattering cross section per unit volume; nonabsorbing component of $\alpha_{aer}(R)$ ($\propto \lambda^{-\gamma}$)

$\sigma_{mol}(R)$ — molecular scattering cross section per unit volume; nonabsorbing component of $\alpha_{mol}(R)$ ($\propto \lambda^{-4}$)

δ — combination of wavelength dependence parameters γ and ε ($\delta=\gamma+\varepsilon$)

Given a homogeneous atmosphere such that $S_\lambda(R)$ is independent of range, then $S_\lambda(R_1) \cong S_\lambda(R_2)$, and the DIAL backscatter error reduces to zero. The range and wavelength dependence of the backscatter mixing ratio must be known when DIAL measurements are attempted in markedly inhomogeneous atmospheres. In that case, steps are required to correct for the large errors that can result from a high degree of aerosol spatial inhomogeneities. As is the case for the extinction error, the magnitude of the backscatter error will be reduced as the wavelength separation between the on and off wavelengths is reduced.

Lidar data can be inverted to provide information on aerosol backscatter and extinction profiles. Analytical solutions to the lidar equation which derive information on aerosol optical properties from independent lidar signals have been reported[11-13]. However, the application of these solutions requires the use of approximations regarding aerosol absorption and backscattering, and they also require additional information on aerosol properties at some calibration range[13,14]. For lidar studies of stratospheric aerosols, this calibration information is typically provided through the identification of "clean" atmospheric regions where aerosol backscattering is sufficiently small[15]. Under the condition of spatial homogeneity, the so-called "slope" method can be applied to the lidar equation to derive the total optical extinction coefficient as follows:

$$\alpha_\lambda = \frac{1}{2} \frac{d}{dR} \ln \left| P_\lambda(R)R^2 \right| \qquad (5)$$

The Bernoulli solution to the lidar equation has been applied to the derivation of aerosol scattering profiles from lidar data[5,13]. Details of the Bernoulli solution as it is applied to the DIAL O_3 correction can be found in Ref. 5. In this approach, the off-line DIAL wavelength is used to estimate the aerosol scattering ratio profile. Since the DIAL backscatter correction depends more on the range dependence of the scattering ratio than on the absolute value, it is expected that the Bernoulli solution will provide estimates for that correction to within ± 5 ppbv for O_3. The sensitivities of the backscatter correction to aerosol optical characteristics and calibration range assumptions are discussed in detail by Browell et al. (see Ref. 5).

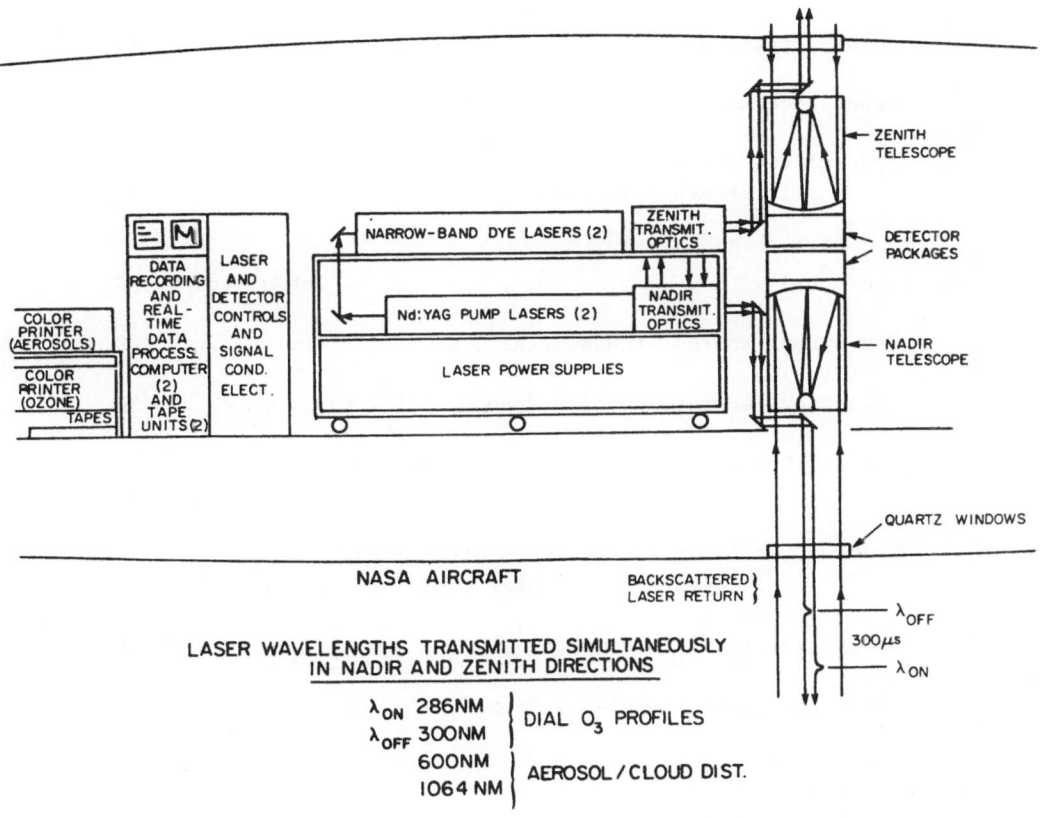

Fig. 2. Schematic of NASA Langley Research Center's airborne DIAL system.

III. NASA AIRBORNE DIAL SYSTEM

The initial development of the NASA Langley Research Center airborne DIAL system was completed in 1980[16], and since that time, many modifications and improvements have been made to this system. A block diagram of the current airborne DIAL system is shown in Figure 2. Two frequency-doubled Nd:YAG lasers are used to pump two high-conversion-efficiency, frequency-doubled, tunable dye lasers. The four lasers are mounted on a structure that supports all of the laser power supplies, the laser beam transmitting optics, and the dual telescope and detector packages for simultaneous nadir and zenith O_3 and aerosol measurements. One of the frequency-doubled dye lasers is operated at 286 nm for the DIAL on-line wavelength of O_3, and the other one is operated at 300 nm for the off-line wavelength. The

Table 1. Airborne DIAL Transmitter Characteristics

Pump lasers (2)	Quantel Model YG 482
Pulse separation, μs	300
Pulse energy at 532 nm, mJ	400
Pulse length, ns	15
Repetition rate, Hz	5 or 10
Transmitted laser energy at 1064 nm (each direction), mJ	100
Dye lasers (2)	Jobin-Yvon HP-HR
Dye output energy, fundamental[1], mJ	120
Dye output energy, doubled fundamental[2], mJ.	30
Transmitted laser energy (each direction), fundamental[1], mJ	35
Transmitted laser energy (each direction), doubled fundamental[2], mJ	15
Laser linewidth, fundamental[1], pm	<8
Laser linewidth, doubled fundamental[2] , pm	<4

(1) Near 600 nm
(2) Near 300 nm (UV)

DIAL wavelengths are produced in sequential laser pulses with a time separation of ≤ 300 s to ensure that the same atmospheric scattering volume is sampled at both wavelengths during the DIAL measurement. Half of each UV beam is transmitted in the zenith and nadir directions. The fundamental dye laser output that is left after the frequency-doubling process and the residual 1064 nm output from the frequency-doubled Nd:YAG laser are also transmitted for aerosol profile measurements. The output laser beams are separated and steered with dielectric-coated optics. They are then transmitted out of the aircraft coaxially with the receiver telescopes through 40-cm diameter quartz windows. Parameters for the airborne DIAL transmitter system are given in Table 1.

The receiver system consists of two back-to-back 36-cm diameter Celestron telescopes with custom made quartz Schmidt corrector plates and two identical detector packages. Each detector package has dichroic optics to separate the UV (286 and 300 nm), visible (600 or 572 nm), and IR (1064 nm) lidar returns onto their respective detectors. Gateable photomultiplier tubes (PMT) are used for the UV and visible returns, and an avalanche photodiode (APD) is used for the IR return. The UV PMT detects the on- and off-line returns sequentially while the visible PMT and IR APD detect one lidar return in each channel. Table 2 gives the characteristics of the receiver system. Each analog signal is amplified at least 10 times prior to digitization by a Transiac Model 2012 transient digitizer, and the lidar signals are digitized at 10 MHz to 12-bit accuracy. The digital data are handled by two LSI-11/73 computers with the raw data being stored on 1600-bpi digital magnetic tape. Two tape units are used to avoid data gaps. Ozone concentrations and aerosol distributions can be calculated simultaneously in real-time, and the output can be displayed on a video screen or can be continuously plotted in color with two ink-jet plotters for in-flight and post mission analysis.

Table 2. Airborne DIAL Receiver Characteristics.

	Wavelength Regions		
	286-300 nm	572-600 nm	1064 nm
Area of receiver, m^2	0.086	0.086	0.086
Receiver efficiency to detector[1], % .	31	40	31
Detector quantum efficiency, %	21 (PMT)	7.2 (PMT)	40 (ADP)
Total receiver efficiency, %	6.5	2.9	12.4
Receiver field of view (selectable), mrad	≤1.5	≤1.5	≤1.5

(1) Includes filter transmission for daytime operation.

IV. APPLICATIONS OF AIRBORNE DIAL TO INVESTIGATIONS OF OZONE AND AEROSOLS

The NASA Research Center airborne DIAL system[16] has been used to investigate O_3 and aerosol distributions in many atmospheric studies. This section discusses examples of the wide variety of O_3 and aerosol measurement made in the troposphere and stratosphere with this DIAL system.

IV.1. Mixed Layer Investigations over Eastern United States

Subsequent to the first airborne DIAL measurements of O_3 profiles in early 1980[16], this system was flown in an EPA field experiment to study Persistent Elevated Pollution Episodes (PEPE) in the summer of 1980[17]. Remote measurements of O_3 and aerosol profiles were obtained on long-range flights over the eastern United States. Enhanced concentrations of O_3 were found in the mixed layer (ML) which resulted from the photochemical production of O_3 during period of atmospheric stagnation with sources of hydrocarbons and nitrogen oxide compounds and high solar insolation. Over large regions during the PEPE experiment O_3 mixing ratios of $\geq$ 90 ppbv were found in the ML while background levels in the free troposphere were only 40 ppbv. Since the major natural source of tropospheric O_3 is from the downward transport of air from the upper troposphere or lower stratosphere, the only possible explanation for having more O_3 in the ML than in the free troposphere is photochemical production of O_3. Ozone in the ML can be removed at the surface by reactions on vegetation and by vertical transport into the free troposphere via cloud dynamics.

A study to make the first direct observations of the cloud pumping of O_3 from the ML into the free tropo-sphere was conducted with the airborne DIAL system in the summer of 1981[18]. The field experiment was designed to follow and characterize O_3 distributions in an air mass that has experienced active and penetrative cumulus clouds. The first measurements were made on July 22, 1981, over the Raleigh/Durham airport during the afternoon in the presence of active convective mixing by cumulus clouds, and a second mission ws flown that evening just west of the North Carolina coast, which was the projected trajectory position of the same air mass sampled in the afternoon. Observations with the airborne DIAL system on the evening flights revealed regions with enhanced O_3 and aerosol concentrations in the free troposphere. These regions were found to contain the remnants of the venting cumulus clouds that were present in the afternoon, and the O_3 mixing ratios in these regions were comparable to the higher O_3 levels (>80 ppbv) in the afternoon ML. These observations present the first evidence for cloud venting of ML O_3 to the free troposphere. This process can be a significant sink for O_3 in the ML, and it must be taken into account when modeling the budget of O_3 in the lower atmosphere.

IV.2. Observations of a Tropopause Fold Event

A tropopause fold event (TFE) is an important phenomenon in stratosphere-troposphere exchange processes. A deformation of the tropopause can occur in the vicinity of the jet stream with the effect that some of the stratospheric air is transported in a layer which is "folded" into the troposphere. During April 1984, the airborne DIAL system obtained the first detailed cross section of O_3 and aerosols across a TFE[19]. Figure 3 shows the data obtained by the DIAL system while operating in a nadir mode from the WFF Electra aircraft at an altitude of 8.2 km ASL (above sea level). Color plots are given for the relative aerosol backscattering distribution (top) obtained at 1064 nm and for O_3 mixing ratio distributions (bottom) obtained from DIAL measurements at 286 and 300 nm. The color scales used are given at the bottom of each color plot. At the top of Figure 3 is a false-color display of the relative aerosol distribution along the atmospheric cross section beneath the aircraft. This display was generated from the airborne lidar data at 1064 nm, with each vertical line representing the average of 15 lidar returns. The black region at the bottom of the aerosol plot is the topography along the Electra flight track. Immediately above the topography, the ML over the desert is readily identified by the enhanced backscatter extending to 3-4.2 km ASL. Cumulus clouds are also seen to be present near the top of the ML. A few of these clouds are optically thick to the laser beam, which results in a shadow beneath them. The free troposphere typically has less aerosol loading than the ML, and this is shown in Figure 3 by the greenish-blue to light blue false coloring. However, embedded in this relatively clean air is a 2 km deep layer with enhanced aerosol loading indicated by yellow and orange that sloped downward from north of Las Vegas to the top of the ML near Yuma. This layer coincides with a layer of enhanced O_3 mixing ratio, as seen in the plot at the bottom of Figure 3. The decrease in O_3 mixing ratio in the layer from the upper right (>240 ppbv) to the lower left (>150 ppbv) conclusively identifies this layer as a tropopause fold. The enhanced aerosol scattering in the stratospheric air was attributed to aerosols from the El Chichon eruption. This was the first complete tracking of an air mass in a TFE from the stratosphere to the top of the ML.

IV.3. Ozone Measurements over Amazon Rain Forest

Rain forests comprise a large fraction of the land areas in tropical regions of the world, and since the large-scale atmospheric motions generated in the Tropics drive much of the tropospheric dynamics at higher latitudes, the unique chemistry associated with the Tropics can have global consequences. A major program to study the chemistry of the lower troposphere over a major rain forest was conducted in the Amazon Basin of Brazil during the dry season of July-August 1985 and the wet season of April-May 1987 as part of the GTE-Amazon Boundary Layer Experiment (ABLE-2A in 1985 and ABLE-2B in 1987). The airborne DIAL system was used to investigate the vertical and horizontal variability of O and aerosols over the tropical rain forest, and these measurements were then related to the sources and sinks for O_3 and aerosols[20,21].

During the 1985 ABLE-2A dry season field experiment, the airborne DIAL system was operated in the nadir mode from the NASA WFF Electra aircraft on 15 flights spanning different areas over the Amazon Basin between Tabatinga and Belem, Brazil[20]. A typical cross section of aerosol and O_3 distributions observed during undisturbed daytime conditions over the rain forest is shown in Figure 4. The relative amount of aerosol backscattering (top) at 1064 nm is defined in the color scale given at the top of the aerosol display, and the scale for the O_3 mixing ratio distribution (bottom) is given at the top of the O_3 display. Universal time (UT) is given at the top of each display (UT is 4 hours ahead of local standard time for Manaus, Brazil). Geometric altitudes are given in kilometers above ground level (AGL). The latitude and longitude are given at the bottom of the plots (e.g., 0604.7/5857.8 is 6°4.7'S latitude and 58°57.8'W longitude). Measurements were made from an aircraft altitude of 4.7 km AGL. False color displays of the relative aerosol-backscattering and O_3 mixing ratio distributions are shown along the aircraft flight track. These displays were generated from airborne lidar data obtained from an altitude of 4.7 km ASL. The top display is the aerosol distribution derived from the 1064-nm lidar channel, with each vertical line representing the average of 15 lidar returns. The top of the forest canopy can be seen as a black line at the bottom of the display. The ML can be readily identified by the enhanced backscattering that extends above the topography. As was noted before, the ML normally contains a naturally higher concentration of aerosols compared to the regions above it. Convective plumes resulting from surface heating effects can be seen at the top of the ML, and there is a noticeable variation in the height of the ML from point B to point A. The black areas at the top of the ML are clouds in various stages of development. The clouds that are optically thick for the lidar have shadows (shown as white areas) beneath them. Some of the clouds have enough buoyancy for them to penetrate the inversion at the top of the ML and ascend to a higher stable level. The area that is predominantly blue in the aerosol display is the old planetary boundary layer (PBL). This is the region that underwent active mixing by clouds on the preceding day. The trade wind inversion (TWI) height can be seen at the altitude where the PBL aerosol loading decreases abruptly to about 40 percent of the value in the PBL. The average height of the TWI is estimated to be about 3.2 km AGL (above ground level).

The O_3 mixing ratio distribution shown at the bottom of Figure 4 was derived from airborne DIAL measurements. Each vertical line represents an average of 300 lidar returns (1-min average or about 7-km horizontal distance) with a DIAL measurement resolution of 210 m. A horizontal running average was used to present a continuous O_3 plot that was in phase with the aerosol data. The start of the lidar data begins about 750 m below and above the aircraft, and because of the 210-m vertical smoothing of the DIAL data, O_3 calculations are not made within about 300 m of the surface. White areas in the O_3 data represent data gaps due to clouds, when they limit the O_3 retrieval to less than 50 percent of the horizontal averaging interval. The O_3 distribution across the leg is horizontally homogeneous, with lower O_3 mixing ratios in the ML compared to those in the PBL, which are also lower than those in the free troposphere. Above the TWI, the O_3, levels increased rapidly into the troposphere. This trend in the O_3 distribution is negatively correlated with the aerosol distribution and supports the theory that in the undisturbed and unpolluted atmosphere over the tropical rain forest, the predominant source of O_3 in the ML and PBL results from downward transport of O_3 from the free troposphere, with destruction of O_3 at the forest canopy.

Large-scale variations in the vertical and horizontal distributions of

O_3 and aerosols were observed on all flights over the Amazon Basin during the dry season. Many regions to the east of Manaus, Brazil, had high levels of O_3 (> 50 ppbv) that were positively correlated with aerosols from biomass burning. A 59 percent increase in O_3 beneath the TWI was found between Manaus (60°W latitude) and Belem (48°W latitude) over a 16-day period due to an increase in biomass burning activity as the dry season progressed into August. The results of the airborne DIAL measurements of O_3 and aerosols during ABLE - 2A[20] and 2B[20] have provided the first extensive data sets for the characterization of the background conditions in the lower troposphere over this important ecosystem.

IV.4. Ozone Hole Measurements.

A joint field experiment between NASA and NOAA was conducted during August-September 1987 to obtain in situ and remote measurements of key gases and aerosols from aircraft platforms during the formation of the O_3 hole over Antarctica. During this experiment the airborne DIAL system was operated in a zenith mode from the NASA Ames Research Center DC-8 aircraft to obtain profiles of O_3 and aerosols in the lower stratosphere on long-range flights over Antarctica[21].

The DIAL system was configured to transmit simultaneously four laser wavelengths (301, 311, 600 and 1064 nm) above the DC-8 for DIAL measurements of O_3 profiles from 10 to 19 km ASL and multiple wavelength aerosol backscatter measurements from 10 to 24 km ASL. A total of 13 DC-8 flights over Antarctica were conducted from August 28 to September 29, 1987. Polar stratospheric clouds were detected in thin layers up to an altitude of about 21 km ASL and were correlated with regions of low temperatures. Large-scale cross sections of O_3 distributions were obtained between 11-20 km ASL inside and outside the polar vortex with a vertical and horizontal resolution of 500 m and 60 km, respectively. Trends seen in the airborne DIAL data compare well with the trends in the O_3 column abundance data obtained from the TOMS (Total Ozone Mapping System) satellite instrument. Figure 5 shows an example of the airborne DIAL measurement of the O_3 depletion over Antarctica on September 26, 1987. The primary altitude region of the O_3 reduction was found to be between 15-22 km ASL. The higher O_3 mixing ratios of >2500 ppbv on the left of the plot are outside the polar vortex, which is at about 62°S latitude, while the region of the O_3 depletion on the right of the plot is inside the polar vortex. The O_3 mixing ratio in the depletion region has a value of <1500 ppbv. The average O_3 concentration at high latitudes (>76°S) in the altitude range of 15-20 km ASL decreased by more than 50 percent between late August to late September. The data obtained by the airborne DIAL system during this field experiment provided the first information on the large-scale vertical variability of O_3 and aerosol profiles during the formation of the O_3 hole over Antarctica.

V. CONCLUDING REMARKS

This paper has discussed the Differential Absorption Lidar (DIAL) technique as it is applied to the measurement of O_3 and aerosols in the troposphere and lower stratosphere. The airborne DIAL system developed at the NASA Langley Research Center was described as an example of an advanced field system that has been used in over 12 major field experiments from 1980-1989. The breadth of the O_3 investigations conducted with this system was presented in this paper with examples of DIAL O_3 and aerosol data from several major field experiments.

The demonstrated accuracy of the DIAL technique for making measurements of O_3 profiles in the troposphere and stratosphere and the recent application of DIAL system in many major atmospheric experiments have generated enthusiasm for the important contributions that DIAL O_3 measurements can make to atmospheric science investigations. As a result, DIAL O_3 measurements are being included as an integral part of nearly all of the future atmospheric studies that deal with O_3 chemistry and transport in the troposphere and stratosphere. This is the first laser remote sensing technique for the measurement of a gas to reach this important degree of maturity.

VI. ACKNOWLEDGEMENTS

The author thanks his colleagues in the Lidar Applications Group for their assistance in the development of the NASA airborne DIAL system, in conducting the field experiments, and in reducing and analyzing the DIAL data presented in this paper. He also acknowledges the assistance of Joy Duke in the preparation of this manuscript.

REFERENCES

1. R. M. Schotland, Some observations of the vertical profile of water vapor by means of a laser optical radar, in: "Proc. of the Fourth Symposium on Remote Sensing of Environment, Rev. Edition", 4864-11-X, Willow Run Lab., Inst. Sci. Technol., Univ. of Michigan, 273-283 (1966).

2. R. L. Byer and M. Garbuny, Appl. Opt., 12:1946 (1973)

3. R. M. Schotland, J. Appl. Meteorol., 13:71 (1974)

4. R. M. Measures, ed., "Laser Remote Sensing, Fundamentals and Applications", Wiley and Sons, Toronto, Canada (1984)

5. E. V. Browell, S. Ismail and S. T. Shipley, Appl. Opt., 24:2827 (1985)

6. T. Kobayashi, Techniques for laser remote sensing of the environment, in: "Remote Sensing Reviews, Volume 3", Harwood Academic Publishers GmbH, 1-56 (1987)

7. M. L. Wright, E. K. Proctor, L. S. Gasiorek and E. M. Liston, A preliminary study of air-pollution measurements by active remote-sensing techniques, NASA CR-132724 (1975)

8. E. E. Remsberg and L. L. Gordley, Appl. Opt., 17:624 (1978)

9. R. T. Thompson, Jr., Differential absorption and scattering LIDAR sensitivity predictions by least squares with application to shuttle, aircraft and ground based systems, NASA CR-2627 (1976)

10. E. P. Shettle and R. W. Fenn, Models for the aerosols of the lower atmosphere and the effects of humidity variations on their optical properties, AFGL-TR-79-0214, U.S. Air Force, Setp. (1979)

11. R. T. H. Collis and P. B. Russel, Lidar measurements of particles and gases by elastic backscattering and differential absorption, in: "Laser Monitoring of the Atmosphere", E. D. Hinkley, ed., Springer-Verlag, 71-151 (1976)

12. F. G. Fernald, B. M. Hermann and J. A. Reagan, J. Appl. Meteorol., 11:482-489 (1972)

13. J. D. Klett, Appl. Opt., 20:211 (1981)

14. W. Hitschfeld and J. Bordan, J. Meteorol., 11:58 (1954)

15. P. B. Russell, T. J. Swissler and M. P. McCormick, Appl. Opt., 18:3783 (1979)

16. E. V. Browell, A. F. Carter, S. T. Shipley, R. J. Allen, C. F. Butler, M. N. Mayo, J. H. Siviter, Jr and W. M. Hall, Appl. Opt., 22:522 (1983)

17. E. V. Browell, S. T. Shipley, C. F. Butler and S. Ismail, Airborne lidar measurements of aerosols, mixed layer heights, and ozone during the 1980 PEPE/NEROS summer field experiment, NASA RP-1143 (1985)

18. J. K. S. Ching, S. T. Shipley and E. V. Browell, Atmos. Environ., 22:225 (1988)

19. E. V. Browell, E. F. Danielsen, S. Ismail, G. L. Gregory and S. M. Beck, J. Geophys. Res., 92:2112 (1987)

20. E. V. Browell, G. L. Gregory, R. C. Harriss and V. W. H. Kirchhoff, J. Geophys. Res., 93:1431 (1988)

21. E. V. Browell, L. R. Poole, M. P. McCormick, S. Ismail, C. F. Butler, S. A. Kooi, M. M. Szedlmayer, R. Jones, A. Kruger and A. Tuck, Large-scale variations in ozone and polar stratospheric clouds measured with airborne lidar during formation of the 1987 ozone hole over Antarctica, in: "Polar Ozone Workshop-Abstracts", NASA CP-10014, 61-64 (1988)

DETERMINATION OF ATMOSPHERIC OZONE PROFILES AT 68N AND 79N

WITH A DAYLIGHT LIDAR INSTRUMENT

R. Neuber

Alfred-Wegener-Institute for Polar and Marine Research
Bremerhaven, Federal Republic of Germany

I. INTRODUCTION

The stratospheric ozone layer has recently received increased attention due to the globally observed decrease of the total ozone concentration[1] and the appearance of the so called ozone hole over Antarctica. Most
of the information on global ozone is based on measurements of the ozone
column density of the atmosphere. The vertical distribution of ozone can
be measured by a light radar or lidar (LIght Detection And Ranging). Laser-
produced short light pulses are sent vertically into the atmosphere. The
light is backscattered by the air molecules (Rayleigh scattering) and the
signals of each pulse are recorded time-resolved, thereby providing the
altitude resolution. For measurements of the ozone concentration the differential absorption principle is used (DIAL = DIfferential Absorption
Lidar): light pulses at two wavelengths are emitted, which are chosen
such, that one is partially absorbed by the ozone molecules and the other
not. But the wavelength difference should be small enough, so that both
waves experience the same backscattering process. Then many unknowns in
the lidar equation, like atmospheric and instrumental parameters, can be
eliminated. A recent description of the application of this principle to
lidar ozone measurements is given by Steinbrecht et al.[2].

In this chapter a lidar instrument is described, which uses the
differential absorption principle to measure atmospheric ozone profiles. A
low bandwidth emitter and receiver allow measurements during daytime.
Results from two Arctic sites are presented.

II. THEORY OF THE OZONE LIDAR

If only single scattering processes occur, the backscattered intensities, which are detected by the receiver of a lidar instrument, are
given as

$$I(R,\lambda) = I_0(\lambda) \ \Delta R \ F/4\pi R^2 \ \varepsilon(\lambda) \ G(R) \ T^2(R,\lambda) \ \beta(R,\lambda) + U$$

Optoelectronics for Environmental Science, Edited by S. Martellucci and
A.N. Chester, Plenum Press, New York, 1990

with I_0 the emitted intensity, R the altitude, F the receiving area, ε the detection efficiency, and G the geometric overlap between the laser beam and the telescope field of view. T denotes the transmittance of the atmosphere, β the backscattering coefficient, and U the background signal (noise and daylight). U will be neglected in the following treatment. As the light can be scattered by the air molecules and by aerosols, $\beta = \beta_{Ray} + \beta_{Mie}$. The transmission term can be written as

$$T(R,\lambda) = \exp(-\int \alpha \, dr)$$

with the extinction coefficient

$$\alpha(r,\lambda) = \alpha_{Ray}(r,\lambda) + \alpha_{Mie}(r,\lambda) + \alpha_{O_3}(r,\lambda)$$

$$= n_{Ray}(r) \, \sigma_{Ray}(\lambda) + n_{Mie}(r) \, \sigma_{Mie}(\lambda) + n_{O_3}(r) \, \sigma_{O_3}(\lambda)$$

which contains the information about the ozone concentration profile $n_{O_3}(r)$. σ is the extinction cross section.

Using the DIAL principle means that (at least) two backscattering profiles $I_1(R,\lambda_1)$ and $I_2(R,\lambda_2)$ are employed, which can be treated as follows. In the ratio I_1/I_2 the instrumental parameters ΔR, F, and G are eliminated. By taking the logarithm of the ratio one gets rid of the exponential in the transmittance term. Then the derivative with respect to R can be taken with two effects: the altitude independent parameters I_0, $\varepsilon(\lambda_1)$ and $\varepsilon(\lambda_2)$ disappear, as does the integral of the transmittance term. Then we obtain for the ozone concentration

$$n_{O_3}(R) = 1/2 \, \Delta\sigma_{O_3} \{d/dR(\ln(I_1(R,\lambda_1)/I_2(R,\lambda_2)))-A(R)-B(R)\} \qquad (1)$$

with $A(R) = n_{Ray} \, \Delta\sigma_{Ray} + n_{Mie} \, \Delta\sigma_{Mie}$,

$$B(R) = d/dR \, \ln(\beta(R,\lambda_1)/\beta_{Ray}(R,\lambda_1) \cdot \beta_{Ray}(R,\lambda_2)/\beta(R,\lambda_2)),$$

$$\Delta\sigma_{O_3} = \sigma_{O_3}(\lambda_1) - \sigma_{O_3}(\lambda_2).$$

The differential backscatter term B(R) depends on the aerosol load of the atmosphere. For low aerosol concentrations (normal back-ground aerosol), it can be neglected. The differential extinction term A(R) contains the extinction of the light pulse due to Rayleigh and Mie scattering. Again the Mie scattering can be neglected for normal aerosol loadings and the Rayleigh extinction can be calculated from an air density profile. If Mie scattering becomes large, as for volcanic aerosols or polar stratospheric clouds, the backscattered intensities of those altitudes can significantly exceed the Rayleigh signal. The aerosol signal can be deduced from the reference wavelength, which is not affected by ozone absorption. For the actual data evaluation the differential d/dR is approximated by the differences between two successive ΔR values. Thus the ozone concentration can be calculated independently from instrumental parameters and most of the atmospheric parameters, but it is essential to know the difference between the two wavelengths in the ozone extinction cross section $\Delta\sigma_{O_3}$. As the ozone extinction cross section is temperature dependent[3], the atmospheric temperature profile also has to be known. Then the main contribu-

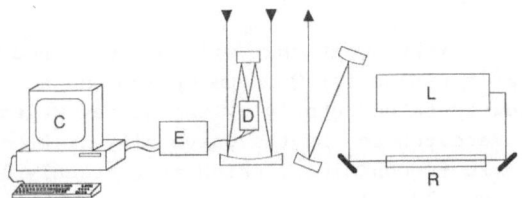

Fig. 1. Schematic of the ozone lidar set up,
with laser (L), Raman cell (R), detector
(D, see figure 2), electronics (E), and
the computer (C).

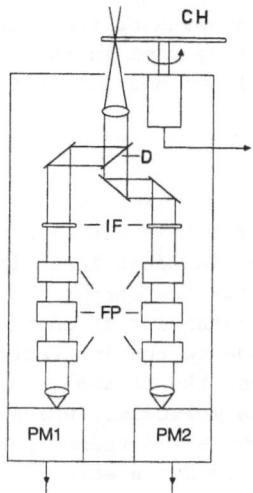

Fig. 2. The detector block,
containing the rotating
chopper (CH), the
dichroic mirror (D),
two interference filters
(IF), up to six Fabry-
Perot etalons, and two
photomultipliers (PM).

tion to the error of the obtained ozone concentrations comes from the Poisson statistics of the backscattered count rates $I(R,\lambda)$.

III. THE INSTRUMENT

The ozone lidar instrument that will be described here was developed at the Max-Planck-Institute for Quantum Optics and at the University of Munich[2]. It is now operated jointly by the Alfred-Wegener-Institute for Polar and Marine Research and by the University of Bremen. The instrument is installed in a 20 ft container, which is currently located at Ny-Alesund, Spitsbergen (79N, 12E).

The lidar, which is shown schematically in Figure 1, consists of three main parts: the laser, the transmitting and receiving optics, and the data acquisition system. Details of the instrument can be found in Table 1.

III.1. The Laser

The laser used is a XeCl-excimer gas laser. A high voltage discharge produces excited dimers of XeCl. As this molecule is unstable in the ground state, the laser conversion is readily achieved. The emitted wavelengths correspond to two transitions with peak intensities near 308 nm, which lie conveniently on the slope of the Huggins ozone absorption band. The Lambda Physik EMG 150T MSC laser consists of two stages: an oscillator and an amplifier. A reflective grating used in the oscillator reduces the bandwidth of the laserlight to 10 pm and allows tuning of the wavelength between 307.7 and 308.4 nm.

III.2. The Raman Shifter

The second wavelength for this DIAL instrument is provided by a hydrogen gas cell, where the incident laser light is partially shifted in wavelength by stimulated Raman scattering. As the incident light intensity is high, a relatively high intensity of spontaneously Raman scattered light is produced. This leads to the interaction of two waves with the gas molecules, the laser wave and the Stokes wave. The two waves are coupled by the excited states of the molecules, which allow a rapid energy exchange between the laser wave and those Stokes waves, which travel in the same direction (stimulated Raman effect). The linewidth of the spontaneous as well as of the stimulated Raman scattered ligth depends on the linewidth of the incident light and is as small as that. Further advantages of a Raman shifter are the high intensity of the shifted wavelength (here up to 40% of the incident laser light), and the fact that both wavelengths are emitted simultaneously and collinearly. (For a detailed description of stimulated Raman scattering see, for example, Bloembergen[4].

In the setup of this instrument, the Raman cell consists of a 150 cm long stainless steel tube, equipped with quartz windows. The incident light is focused by a f = 100 cm lens and defocused after the tube by a f = 62 cm lens. The gas used is hydrogen at pressures of about 2 bar, which yields conversion rates of up to 40%. The shifted wavelength is 353 nm, which is away from any strong ozone absorption band.

Table. 1. Parameters of the Lidar Instrument.

The Laser:

model	: Lambda Physik EMG 150T MSC
type	: XeCl-excimer gas laser
wavelength	: 308 nm (tunable 307.7 - 308.4 nm)
bandwidth	: 10 pm
energy	: 150 mJ/pulse
pulselength	: 15 ns
repetition rate	: 40 Hz (max.)
	25 Hz (nominally)
divergence	: < 0.2 mrad
raman shifter	: hydrogen gas tube,
	150 cm long,
	2 bar nominal pressure

The Receiver:

mirror diameter	: 60 cm
focal length	: 240 cm
field of view	: 0.15 mrad - 0,3 mrad
bandwidth	: < 20 pm
filters	
per wavelength	: 1 interference filter (FWHM = 4 nm)
	2-3 pressure tuned Fabry-Perot etalons
altitude resolution	: 200 m to 3000 m
altitude range	: 12 to 35 km in daylight
	12 to 50 km during night
integration time	: 5 h for daylight measurements
	2 h for night measurements (typically)

III.3. The Sending and Receiving Optics

The emitted laser light is sent vertically into the atmosphere by a Galilei telescope consisting of a convex and a concave mirror. It acts as a beam expander, which reduces the divergence of the laser beam by a factor 10. A low divergence is necessary to keep the laser beam inside the field of view (FOV) of the receiving telescope. The FOV should be as small as possible to reduce the background light, which is seen by the receiver. The axis of the sending and receiving optics are separated by 50 cm; therefore the laser beam and the FOV overlap completely a few kilometers above the container. The receiving optics consist of a concave mirror of 60 cm diameter and a focal length of 240 cm. The focus is folded into the detector by a plane mirror (see Figure 1).

III.4. The Detector

In Figure 2 a schematic diagram of the detector is shown. A mechanical chopper blocks out the high intensity light, which is backscattered from the lower altitudes. This chopper also provides the timing signals for the firing of the laser and the start of the data acquisition for each laser pulse. A dichroic mirror is used to separate the two backscattered wavelengths. The bandwidth for each channel is reduced by an interference filter and up to three pressure-tuned Fabry-Perot etalons. This is possible without great losses in signal strength, as the emitted laser light is also of small bandwidth. Therefore most of the background day-light and any light from the other wavelength is excluded. The free spectral range of the Fabry-Perot etalons differs only slightly, to allow only one combined transition maximum under the transmittance curve of the interference filter. The etalons cannot be tilted, but are adjusted inde-pendently by changing the pressure of the SF_6 gas inside each. A temper-ature stabilization secures stability for long periods (up to some weeks). Below the etalons the light is focused into Thorn EMI photomultipliers, which work in the photon counting mode.

III.5. The Data Acquisition System

The electrical signals from the photomultipliers are transformed to countable pulses by two discriminators. A dedicated double 256-channel counter collects the pulses into time "slots", which correspond to an altitude resolution of 200 m and an altitude range from 0 to 51.2 km. After each laser pulse the collected data from the counter is transferred to the PDP11 microcomputer. The data acquisition system works at a fre-quency of 50 Hz, but the laser is actually fired only with 25 Hz. So with every second data sampling the background signal is recorded. After 1000 laser pulses, the background in each ΔR is subtracted from the signals and the data are stored on disk together with the measurement time and the average background signal.

III.6. The Data Evaluation

To determine the ozone profile according to equation (1), additional information about the atmospheric temperature and density profile is required. This is obtained from a nearby meteorological station. The ozone concentration is then calculated for each 200 m separately, beginning above the ΔR where the chopper has completely opened the detector. The statistical error of the ozone concentration is also determined. If this value surpasses a preset limit, the raw data are integrated over the next ΔR values until a sufficiently low error is obtained. Then the next ΔR is evaluated in the same manner. Once the error cannot be reduced further or the integrated altitude range becomes too large, the upper end of the detectable ozone profiles has been reached and the evaluation is stopped.

IV. RESULTS

This lidar instrument has collected many datasets during the last two years. It was first used on the German research vessel "Polarstern" during an Arctic and a cross-equatorial cruise[2]. During winter 1987/88 the container was installed at Kiruna, Sweden (68N, 21E)[5]. Since summer 1988

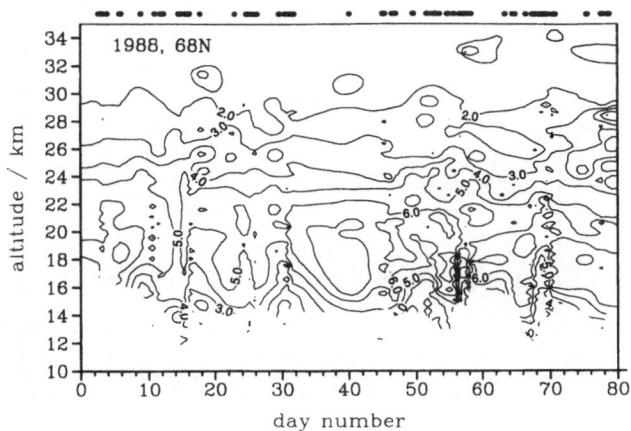

Fig. 3. Plot of isolines of the ozone concentration found above Kiruna, Sweden, from January to March 1988. Values are given in 10^{12} cm^{-3}. The solid bars in the upper row show the times of lidar measurements.

it is located at Ny-Ålesund, Spitsbergen (79N, 12E). The datasets obtained from the last two locations will be briefly described.

IV.1. Comparison of Lidar and ECC - Sonde Data

Besides the lidar a radiosonde system for ECC ozone sondes is operated at Ny-Ålesund to cover extended periods of cloudiness. For comparison several sondes were also launched during lidar measurements. Although the balloons travel with the wind and a lidar profile has to be integrated over several hours, the agreement between both methods is good. Differences range usually below the 10% level and no systematic differences have been found.

IV.2. The Stratospheric Ozone Layer Above Kiruna in Winter 1988

The contour plots of the stratospheric ozone concentration in Figures 3 and 4 show our measurements at Kiruna and Ny Ålesund. In the upper rows solid bars denote the times of lidar measurements and thin bars in Figure 4 the times of sonde launches. In Kiruna the maximum of the ozone concentration was found at about 21 km altitude, with values of $6 \cdot 10^{12}$ cm^{-3} to $7 \cdot 10^{12}$ cm^{-3}. The upper part of the layer is rather stationary, while the lower part displays several short decreases in the ozone concentration. They modulate a general increase starting in mid-February.

IV.3. The Temporal Development of the Ozone Concentration above Spitsbergen

In Figure 4 a contour plot of the ozone layer above Spitsbergen during winter 1989 is shown. During January and the first half of February

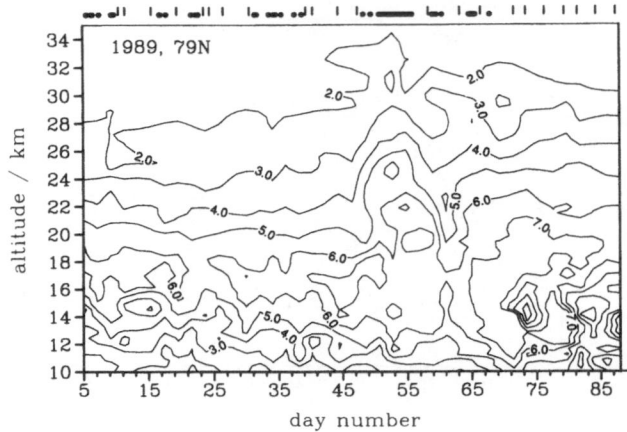

Fig. 4. Plot of isolines of the ozone concentration found
above Spitsbergen from January to March 1989. Values
are given in 10^{12} cm^{-3}. The solid bars in the upper
row show the times of lidar measurements and the
thin vertical lines the times of ECC-sonde launches.

1989 no major changes occurred in the ozone profiles. Especially the upper
part of the ozone layer is quite stable. The altitude of the maximum
concentration descended slowly from 18 km to 16 km, while the maximum
values increased from $6 \cdot 10^{12}$ cm^{-3} to $7 \cdot 10^{12}$ cm^{-3}. The lower part of the
layer is slightly perturbed due to transport processes. The downward slope
of the isolines correlates with the temporal development of the temperature. After mid-February a strong stratospheric warming event occurred
in 1989, which had significant impact on the ozone layer. Two distinct
pulses are found, the first one from 17 Feb. (day 48) to 28 Feb., and the
second one beginning around 6 March (day 65). For the rest of the measurements, until the end of March, the upper part of the ozone layer remained
quite stable. After 6 March (day 65) the ozone layer seems to be split,
with a secondary concentration maximum around 13 km altitude appearing.
This feature is common especially in the mid latitude ozone layer during
spring (see e.g. Krueger and Minzner[5]).

Comparing the data from Kiruna and Spitsbergen, one notes that during
winter the maximum concentration values are similar, but the altitude of
the ozone concentration maximum is found 3 to 4 km lower above Spitsbergen
than above Kiruna. In both cases the upper part of the layer shows little
variation until the breakdown of the polar vortex. (It occurred after mid-
March in 1988; a first hint at increasing concentration values above 25 km
altitude might be found in the last measurements from Kiruna). The lower
part of the layer is much more variable, which could be observed with good
temporal resolution with this lidar.

V. ACKNOWLEDGMENTS

The lidar was operated by B. C. Krüger and the author. We very much appreciate the help and introduction received from W. Steinbrecht. We gratefully acknowledge the support by the staff of ESRANGE, Kiruna, and of the Norsk Polarinstitutt, Ny-Ålesund. This project was financed by the Bundesministerium fur Forschung und Technologie.

REFERENCES

1. Present State of Knowledge of the Upper Atmosphere 1988: An Assessment Report ("Ozone Trend Panel Report"), NASA Reference Publication 1208 (1988)
2. W. Steinbrecht, K. W. Rothe and H. Walther, Appl. Opt., 28, 3616-3624 (1989)
3. R. J. Paur and A. M. Bass, The Ultraviolet Cross Sections of Ozone: II. Results and Temperature Dependence, in: "Atmospheric Ozone", Proc. Quadr. Ozone Symp., C. S. Zerefos and A. Ghazi, eds., Dordrecht (1985)
4. N. Bloembergen, Nonlinear Optics, Benjamin, New York (1977)
5. A. J. Krueger and R. A. Minzner, A Mid-Latitude Ozone Model for the 1976 U.S. Standard Atmosphere, JGR 81, 4477 (1976)

OTHER ATMOSPHERIC SENSING

LASER ATMOSPHERIC WIND VELOCITY MEASUREMENTS

R. T. Menzies

Jet Propulsion Laboratory
California Institute of Technology
Pasadena, California

I. INTRODUCTION

The value and use of wind field measurements in environmental and
Earth system science is discussed as an introduction. A brief overview or
remote sensing techniques which can be used for wind field measurements is
then included to provide a context for evaluating the particular characteristics of Doppler lidar which enhance wind measurement capability. These
techniques include cloud-track winds measurements, aerosol or water vapor
pattern correlation techniques, various limb sounders, and Doppler radar.
Principles of Doppler lidar are then introduced, along with a discussion of
the means of providing a backscatter signal which can be used to detect
wind velocity. The discussion of Doppler lidar principles and techniques
distinguishes coherent and incoherent (direct detection) lidar. For each of
these two classes of Doppler lidar, SNR expressions are developed, and the
rationale behind choice of wavelengths and laser technology is discussed.
Examples of Doppler lidar use in field measurements are cited, and several
design considerations for future use of this technique to measure global
wind fields are presented.

The Doppler lidar technique is being used currently for studies of
atmospheric dynamics in the troposphere in field measurement compaigns
organized by research institutions (e.g., the U.S. National Oceanic and
Atmospheric Administration Environmental Research Laboratory). Its use in
other applications, such as airborne true-air-speed and wind shear detection, is the subject of experimental studies and analytical systems
studies. One application that would place demanding requirements on the
coherent lidar technology and represent a major commitment of resources is
an Earth-orbiting Doppler lidar that measures atmospheric wind fields on a
global scale with sufficient spatial and temporal resolution to provide
data for advanced global weather forecast models. The potential of this
technique for global atmospheric wind measurements has motivated several
studies that attempt to define the instrument parameters required for this
application. Overall efficiency is of utmost importance in this applica-

Optoelectronics for Environmental Science, Edited by S. Martellucci and
A.N. Chester, Plenum Press, New York, 1990

tion, since power, weight, and size are restricted. This paper emphasizes the Doppler lidar technique with the global wind measurement application in mind.

The importance of wind field measurements on a global scale, with altitude resolution of approximately 1 km throughout the troposphere and lower stratosphere, is widely recognized by the community of numerical weather prediction (NWP) investigators. Simulation experiments using the latest NWP models show that forecast skill is significantly improved with the use of such data. Direct wind observational data would circumvent the problems associated with the use of the geostrophic adjustement assumption when deducing motions from pressure data. (Pressure, or geopotential height fields, can be calculated from surface pressure and temperature profile data.) The geostrophic approximation assumes balance between pressure and Coriolis forces; however, at low latitudes, where the Coriolis forces are small, the "equilibration time" is very long, and motions often cannot be predicted using the geostrophic approximation. In addiction, the use of mass (geopotential height) fields to deduce motions requires differentiating the pressure with respect to a horizontal dimension. Differentiation enhances the influence of noise in the observational data and the influence of discontinuities, such as gaps in measurement swaths, etc.

During the past decade several organizations have organized studies and workshops to discuss the feasibility of global atmospheric wind field measurements from Earth orbit. The consensus opinion was that lidar offered the only means of mapping wind fields at high vertical resolution (≤ 1 km) throughout the troposphere. Passive radiometers have the potential for accurate horizontal wind measurements in the stratosphere and mesosphere, but not in troposphere. Instruments such as the microwave limb sounder require a low pressure environment to reach accuracy levels of a few meters per second, which limits their application to upper stratosphere and mesosphere dynamics. Correlation spectrometers and high-resolution interferometers that sense spectral line shifts in emission spectra of scattered sunlight spectra are most sensitive in the limb-viewing mode, which restricts the practical opportunities to view the lower stratosphere and troposphere along a clear line-of-sight. In addition, at mid-tropospheric levels and lower, the lines used for Doppler measurements are influenced by pressure broadening, making the measurement of small shifts much more difficult. Other techniques for sensing tropospheric dynamics from space have serious limitations in regard to the spatial or altitude coverage. Cloud-track wind data have been used for over a decade to infer atmospheric dynamics. Principal limitations are: (1) winds can be measured only at the altitudes and geographical coordinates where trackable clouds exist; and (2) knowledge of the cloud-top heights is required. (Cloud-top temperatures, inferred from radiances measured by multichannel radiometers, are used to assign heights.)

Doppler lidars can employ either coherent or direct (incoherent) detection. An efficient coherent lidar makes optimum use of the unique properties of spatial and temporal coherence of the laser transmitter and local oscillator (LO). This requires single mode oscillation and a high degree of frequency stability from the transmitter, which may be either continuous wave (cw) or pulsed, and the LO, which is usually operated cw. For Doppler measurements of moving objects or atmospheric wind fields, an ideal receiver contains a filter that is matched to the temporal coherence of the return signal. The coherence time of the laser transmitter should be slightly longer than that of the return signal or the coherence time corresponding to the minimum detectable velocity, in order to avoid loss of resolution. For example, a velocity shift of 1 m/s produces a 200 kHz Doppler shift at a 10 μm lidar wavelength, and a signal with a 1/e coherence time of 1.2 μs produces a velocity spread of 1 m/s. This implies a laser signal whose spectral width is approximately 10^{-8} times its frequency. The direct detection Doppler lidar also demands a narrow line-width laser transmitter; however, since the optical filtering does not have the spectral resolution which can be obtained using heterodyne detection followed by RF filtering, a transmitter which is single mode, with a spectral width of approximately 30 MHz in the near uv or visible is adequate. With a receiver spectral resolution of 50 MHz, any further narrowing of the transmitter spectral width would not result in signif- icant performance improvement. The receiver would most likely consist of tandem Fabry-Perot etalons, with prefiltering to narrow the spectrum observed by the first (moderate resolution) etalon (as described by Rees[1]). The moderate resolution etalon is designed to pass a spectral window corresponding to the free-spectral-range (FSR) of the high-resolu- tion etalon. Thus its resolution is in the 500-1500 MHz range. Direct- detection Doppler has received some attention[2,3] as a possibility for space-based global wind field measurement instrument. However the use of Doppler lidar in atmospheric studies to date has taken place predominantly with coherent detection systems. For this reason and for the reason that comparative performance efficiency studies have shown[4] that in the present time frame coherent Doppler lidars have an advantage for space-based wind field measurements, the emphasis of this paper is on coherent Doppler lidar. Much of the material in this paper is treated in greater depth in Menzies and Hardesty[5].

II. COMPARISON WITH DOPPLER RADAR

The coherent Doppler lidar has several similarities to Doppler radar, but is is important to note the unique features of Doppler lidar in this context. Similarities exist particularly with regard to the technique used to process the signal after the downconversion which occurs in the photo- mixer of the lidar and the rf mixer of the radar. Much of the signal processing that has been applied to Doppler lidars in wind field measure- ment applications was essentially developed earlier for Doppler radar. However, there are significant differences between the two when one considers the source of the backscatter signal, coherence time, spatial measurement scales, scan patterns available, and effects of clouds and precipitation. Fundamentally, both radar and lidar wind measurements are formed by irradiating a volume of air with coherent radiation and observ- ing the Doppler shift in the frequency of the radiation backscattered from scatterers within the irradiated volume. Doppler radar measures the Doppler shift by observing the phase of the signal over a sequence of pulses, whereas Doppler lidar measures the Doppler shift from a single pulse.

Sensitivity of the system to scatterers of various size is related to the system wavelength. In the radar case, scatterers are refractive index fluctuations, hydrometeors, windborne insects, seeds, or cloud particles. Because Doppler lidar operating wavelengths are usually 3-4 orders of magnitude shorter than radar wavelengths, the primary source of back-scattered energy is small atmospheric aerosol particles whose diameters are within an order of magnitude of the lidar wavelength. With typical atmospheric aerosol size distributions, most of the energy received by a Doppler lidar system, operating at a wavelength of 10 μm, is scattered by particles width diameters of about 1-3 μm.

Because of their short operating wavelength, Doppler lidar systems possess many unique qualities that complement the capabilities of radar techniques. Meteorological radars operating at wavelengths of 3-10 cm are very useful for measuring winds in the presence of hydrometeors such as rain and snow. Such radars operate only marginally in the clear air, however, requiring the presence of insects, seeds or other convectively lifted phenomena to provide echoes. On the other hand, wind profiling radars, which operate with wavelengths ranging from about 6 m to 30 cm, obtain excellent clear air signals from refractive index fluctuations in the optically clear air. The long operating wavelengths necessitates large antennas, which are not easily scanned. Even with the large antennas, however, ground clutter from the antenna sidelobes prevents these radars from being scanned at low elevation angles.

The capabilities of Doppler lidar do not directly overlap the capabilities of either of the meteorological or the wind-profiling Doppler radars. A lidar can operate in the optically clear air like a wind profiler, yet, because the lidar beam can be tightly collimated, the beam can be directed adjacent to terrain or other features without contamination of the return by energy scattered from the antenna sidelobes. Because of the short wavelength, the large antennas required with radar systems are replaced by small, easily-scanned mirrors. Thus, lidar techniques are ideally suited to studies of wind flow in complex terrain, near the surface, or in the presence of objects such as buildings, bridges, etc. The narrow optical beam produced by lidar (1-2 m in diameter) is also useful for examining the fine structure of winds under highly stratified conditions, such as those in the nocturnal boundary layer.

Doppler lidar range resolution is similar to that of a typical scanning Doppler radar. Although the narrow beam provides sharp transverse resolution (on the order of a few meters), range resolution is much coarser (i.e., on the order of 150-300 m for a CO_2 Doppler lidar operating at 10 μm). Although pulses can be shortened to provide somewhat better resolution, the resulting increase in signal bandwidth eventually degrades the Doppler accuracy of the system to the point where the estimates are of marginal value.

III. PROPERTIES OF COHERENT LIDAR

The coherent Doppler lidar uses a pulsed transmitter having a narrow

spectral width approximately comparable with the Doppler shift expected
from a volume of aerosol particles moving at the minimum detectable
velocity. A phase-coherent photomixing receiver converts the optical or
infrared frequency spectrum of the signal backscattered from the moving
aerosol down to radio frequencies. (This detection process is usually
referred to as heterodyne detection). At this stage radio frequency techniques can be used to amplify and filter the signal prior to the input to
transient digitizers and other hardware for fast Fourier transform (FFT) or
multi-lag complex covariance calculations. Several factors must be considered in order to determine the optimum transmitted pulse duration. It
must be long enough to avoid excessive spectral broadening of the return
signal and the consequent loss of resolution in the frequency (or velocity)
domain. On the other hand, it must be short enough to achieve the desired
range resolution. Laser physics and engineering are also important influences on choice of pulse duration, because overall laser efficiency may be
a strong function of both pulse duration and pulse repetition frequency
(prf).

The spatial coherence requirement in photomixing gives the infrared
heterodyne radiometer antenna properties, in similarity with radio-wave-
length radiometers. The phase fronts of the signal and local oscillator
must match at the photomixer in order to obtain maximum mixing efficiency.
For the heterodyne configuration shown in Fig. 1, the focal spot sizes of
the signal and local oscillator beams at the photomixer should be nearly
equal, and they should overlap in order to achieve optimum mixing
efficiency. In this case, the receiver field of view will be $\Omega_r \cong \lambda^2/D_c^2$,
where λ is the radiation wavelength to which the radiometer is sensitive
and D_C is the diameter of the collecting lens or mirror in the optical
receiver (i.e., the entrance pupil). If the local oscillator spot size on
the photomixer is increased by reducing D_{LO}, the local oscillator beam
diameter prior to focusing onto the photomixer, a larger field of view can
be obtained while maintaining overlap at the photomixer. However, the
mixing efficiency is decreased, for the signal coming from any given
direction within the field of view will mix with only a part of the local
oscillator, while the entire local oscillator power induces shot noise in
the receiver. Siegman[6] demonstrated that for any heterodyne receiver
configuration, the sensitivity of the receiver can be calculated by
assuming it has an integrated effective aperture

$$\int A_e(\Omega)d\Omega \cong \lambda^2 \tag{1}$$

where λ is the wavelength of the radiation to which the receiver is
sensitive. This relation is equivalent to the well-known antenna theorem at
radio frequencies.

The spatial coherence requirement results in the property that, when a
heterodyne radiometer is viewing an extended source of thermal radiation,
the radiometer responds to only one spatial mode of the thermal radiation
field. The thermal radiation source can be described in terms of a complete
set of orthonormal modes. The technique of "back-propagating" the local
oscillator field from the receiver entrance pupil to the far field can be

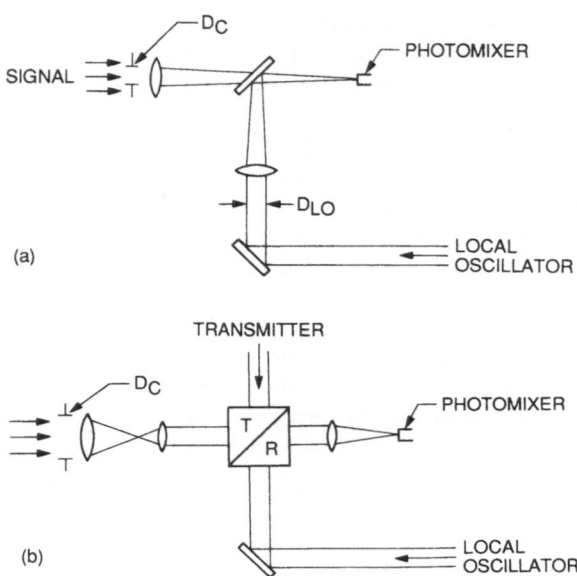

Fig. 1. An optical diagram of heterodyne detection elements in a
passive radiometer, (a); and, analogous optical diagram for
a coherent lidar, with transmit/receive diplexing elements
behind the telescope, (b).

used[6-8] for computing the efficiency of spatial overlap with the signal
source field at a distant plane rather than at the receiver entrance pupil
or at the photomixer surface itself. Calculations of heterodyne mixing
efficiencies assuming parameters which are characteristic of actual
receivers indicate that mixing efficiencies above 0.8 are possible (not
including the photomixer quantum efficiency) when viewing an extended
source of thermal radiation. This is neglecting losses in the optical of
the receiver, one can achieve an efficiency of at least 0.8 η_{QE} compared
with the ideal receiver which responds perfectly to a single mode of the
thermal radiation field, where η_{QE} is the photomixer quantum efficiency.

The backward propagating local oscillator (BPLO) technique has been
used by a number of individuals to calculate heterodyne mixing efficiencies
for various coherent lidar configurations when the signal source is either
a glint target (not spatially resolvable by the receiver optics) or a random
array of point scatters. (See Shapiro[9] and references contained therein.)
The usual optical configuration which is considered in these analyses is a
coaxial configuration in which the transmitter and receiver share a common
optical telescope at the front end. (The telescope is the equivalent to the
collecting, or objective, lens in the simplified example depicted in Fig. 1.
In the case of a coaxial coherent lidar, the LO focusing lens and beam
splitter in Fig. 1 are replaced with transmit/receive diplexer optics.) The
conclusion reached from these analyses is that the overall phase coherent
mixing efficiency of a coherent lidar, including antenna aperture trunca-

tion losses, can be as high as 0.8 in the glint target case, when compared with an ideal lidar with a transmitter antenna gain which is given by the diffraction limit of the front-end telescope and receiver whose antenna cross section is λ^2, as in Eq. (1). However, in the case of a random array of point scatterers (e.g., an extended target whose surface is rough compared with λ), the heterodyne mixing efficiency is less than 0.5. The drop in heterodyne mixing efficiency in the diffuse target case is physically due to the fact that the scattered radiation pattern is a speckle pattern, and when the transmitted intensity profile is adjusted to minimize aperture transaction losses, the resulting backscattered speckle lobe size is such that there is only partial coherence of the field over the receiver aperture. The lower limit of loss, approximately -4 dB, is achieved with a Gaussian profile transmitter whose mode radius (the $1/e^2$ irradiance radius) is approximately 80% of the transmitter (antenna) aperture radius. For other transmitter cross sectional irradiance profiles, such as those produced in the unstable resonator cavities[10] that are suitable for efficient energy extraction in large volume lasers, the mixing efficiency drops somewhat further. The design of unstable resonator cavities that minimize the additional loss of heterodyne mixing efficiency while maintaining high extraction efficiency from the laser gain medium is an active area of research for those involved in the study of an Earth-orbiting Doppler lidar, an application that demands the optimization of overall energy efficiency because of limited available power. Variable reflectance output couplers[11] could be used in various unstable resonator cavity configurations to generate far-field that approach the ideal truncated Gaussian and certain transmitter irradiance patterns that approach a minimum of power in the wasteful higher order lobes. However, it is difficult to efficiently extract the energy near the edges of the discharge in a TEA-CO_2 laser (i.e., near the discharge electrodes) while maintaining a spatial profile with a smooth, Gaussian-like transition from the peak intensity to the edge of the beam profile. Tradeoffs are necessary in the design[12] of the coherent lidar transmitter resonator and telescope optics in order to maximize the overall lidar efficiency.

The lidar received signal power and power signal-to-noise ratio (SNR) are dependent on various instrumental and atmospheric parameters, which must be considered when Doppler lidar is being evaluated for various applications. Consider for example, a pulsed lidar system which responds to aerosol backscatter signals at ranges which are large compared both with $c\tau_p$ (where τ_p is the transmitter pulse duration and c is the speed of light) and with the range gate. Let the transmitted pulse start at t = 0 and end at time t = τ_p with a temporal power profile $P_T(t)$. The received power due to the aerosol backscatter can be expressed as

$$P_b(t) = \int_{c(t-\tau_p)/2}^{ct/2} P_T(t-2R/c)\beta(R)(A/R^2)\eta_h\eta_o O(R)\tau(R)dR, \qquad (2)$$

where A is the receiver area, R is the range, η_h is the heterodyne mixing efficiency, η_o is the system optical efficiency (due to absorption and scattering losses in the optical elements), O(R) is the range-dependent telescope overlap function, $\beta(R)$ is the aerosol volume backscatter coefficient (m^{-1} sr^{-1}), $\tau(R)$ is the two-way transmission along the path to range R at the lidar wavelength, and where the integration over R indicates that the received power at time (t) is due to contributions from a slab of atmosphere of thickness $c\tau_p/2$ centered at $R_b = c(t/2 - \tau_p/4)$. If we assume $c\tau_p$ is smaller than the range gate, and that the dependencies of $\beta(R)$, A/R^2, $O(R)$, and $\tau(R)$ on range over the range-gate interval can be neglected, then, the received signal energy obtained by integrating $P_p(t)$ over the time interval corresponding to the range gate is proportional to the transmitted pulse energy, the aerosol β, the range-gate length, the receiver solid angle (A/R^2), the round-trip atmospheric transmission factor τ, and various system factors.

The ability to measure the frequency of the backscattered signal, and hence the Doppler shift, depends on the ratio of $P_b(t)$ to the noise power of the coherent receiver (the SNR) and on the spectral width of the back-scattered signal. The "short-term" SNR (averaged over a time less than the coherence time of the signal) is given by

$$SNR = \eta_{QE} \, P_b/h\nu B \tag{3}$$

where η_{QE} = the photomixer quantum efficiency, B = receiver bandwidth, and $h\nu$ is the photon energy. This corresponds to a phase coherent, spatially matched photon arrival rate at the photomixer surface of $F = P_b/h\nu$. Although the velocity estimate uncertainty depends in a rather complex way on the SNR, for low values of SNR it is inversely proportional to SNR. Thus, maximization of SNR is of critical importance. The velocity estimate uncertainty and its dependence on SNR and other parameters is discussed in more detail in the following section.

IV. OBTAINING WIND VELOCITY ESTIMATES WITH DOPPLER LIDAR

A Doppler lidar system measures wind field characteristics by irradiating a region of the atmosphere with coherent radiation, then sensing the Doppler shift in the frequency of the radiation backscattered by atmospheric aerosol particles within the irradiated region. Continuous-wave (cw) Doppler lidar systems employ focusing to define the beam dimensions; the range response of such a system is roughly proportional to the inverse of the beam diameter squared. Thus, systems with practical telescopes (1 m diameter or less) have poor range resolution at ranges beyond a few hundred meters. Continuous-wave lidar systems are typically used to probe small atmospheric volumes at ranges close to the system, where focusing can be effectively utilized. In a pulsed laser system, the radial dimension of the instantaneous scattering volume is defined by $c\tau_p/2$, where τ_p is the laser pulse duration; transverse dimensions are determined by the beam dimensions as in the cw case. Thus, by transmitting a pulse and processing the aerosol-backscattered signal as a function of time, range resolved measurements of radial wind speeds are obtained.

When a volume of aerosol particles is irradiated with single frequency laser radiation, each of the particles scatters a portion of the incident radiation back toward the lidar system. Because the particles are randomly positioned within the volume, the phase of the scattered radiation from a single particle is random with respect to that from another particle. The aerosol ensemble, then, acts as an incoherent scatterer. The signal incident on the receiver is formed from the superposition of the radiation backscattered by each individual particle. Since the individual signals have random phase, and the total number of particles is large, the optical signal over a single coherence region at the receiver can be characterized as a single realization of a Rayleigh phasor, with uniformly distributed phase and Rayleigh-distributed amplitude. Said another way, the lidar observes a single lobe of the laser speckle pattern produced by the incoherent aerosol target.

If all of the irradiated particles were stationary with respect to each other, the signal at the detector would consist of a sinusoidal signal of constant random amplitude and phase with frequency equal to that of the transmitted pulse, plus or minus the Doppler shift. However, in reality, the particles move with respect to each other (i.e., move at different velocities) due to turbulence within the scattering volume and wind shear across the volume. Thus, the phase and amplitude of the resultant signal at the receiver change as the relative phase between each particle in the ensemble varies. This change in phase and amplitude reduces the coherence and broadens the bandwidth of the optical signal, such that the shape of the backscattered signal power spectrum is proportional to the distribution of small-scale radial wind speeds within the scattering volume. To a rough approximation, the coherence time t_c of the return due to the ensemble of winds within the volume is estimated by

$$t_c \cong \lambda/\sigma_v$$

where λ is the wavelength and σ_v is the standard deviation of the distribution of radial velocities of the scatters in volume. For a 10 μm lidar, t_c is on the order of a few μs. When pulsed systems are employed, an additional bandwidth-broadening mechanism is introduced due to both laser pulse spectral characteristics and propagation of the pulse through the region of distributed aerosol particles. As the pulse travels, new aerosol particles are continually irradiated; within a time interval τ_p equal to the pulse duration a completely new ensemble of aerosols with a different resultant phase provides the return at the detector. The new resultant signal has no correlation with the signal received τ_p seconds earlier. In an ideal atmosphere, with no shear and turbulence within the scattering volume, the power spectrum of the return would be equal to the power spectrum of the laser pulse. In a real atmosphere shear and turbulence contribute to the bandwidth of the return; the power spectrum of the return is the convolution of the power spectrum resulting from the shear and turbulence, assuming a cw signal, with the power spectrum of the laser transmit pulse.

The radial wind estimate is proportional to the shift of the mean frequency of the aerosol-backscattered signal. Both digital and analog techniques can be employed to estimate the signal mean frequency. CW lidar systems often employ filter banks or surface-acoustic wave (SAW) devices for spectral analysis. Because a signal is always present, the output of the analog process, which is a representation of the spectrum of the signal plus noise current, can be averaged over many coherence time intervals (where coherence time is roughly the reciprocal of the signal bandwidth) to reduce the noise in the spectrum estimate. For example, a 10 μm lidar system might employ a filter bank comprised of filters with 100 kHz bandwidth (this corresponds to about 0.5 m s^{-1} Doppler resolution). Since the time constant of each filter is about 10 μs, 10^5 independent samples can be obtained each second; this yields a significant reduction in the variance of the output of each filter and enables better velocity estimates.

In pulsed Doppler lidar systems, digital processing is usually employed to estimate the mean frequency of the return. Processing can be based on either the complex autocovariance function of the return or on the discrete Fourier transform of time samples. For the case of a Fourier transform analysis of the received signal, Zrnic[13] derived the following expression for the uncertainty of the velocity estimate.

$$\delta v = \frac{\lambda}{4\pi} \left(\frac{f}{T}\right)^{1/2} \left[\pi^{3/2} W + \frac{8\pi^2 W^2}{SNR} + \frac{\pi}{3(SNR)^2}\right]^{1/2} \qquad (4)$$

where λ = the lidar wavelength, f = sampling frequency, which is related to the maximum measurable Doppler shift, or wind velocity as dictated by the Nyquist criterion ($V_{Ny} = f\lambda/2$); T is the integration time; and W is a (normalized) measure of the frequency spread of the return signal in the absence of noise, $W = (V_{Ny})^{-1}(V_{bw}^2 + V_{atm}^2)^{1/2}$, where V_{bw} is the velocity uncertainty due to the pulse bandwidth, and V_{atm} is the standard deviation of the velocity distribution in the measured volume due to turbulent eddies and bulk wind shear effects]. For example, we might assume the following values for the quantities $V_{bw} = \lambda/2\pi t = 0.4$ m/s (for a rectangular pulse of duration t), $V_{atm} = 0.4$ m/s, and then W could be calculated for various Nyquist velocities. (The assumed value V_{atm} is representative of a combination of the midrange between negligible and light turbulence intensity categories.) Caution must be used when applying Eq. (3) at optical and infrared wavelengths for low values of SNR. The SNR used in Eq. (3) is the "wideband" SNR, which is related to the sampling frequency f. It is a certain fraction of the "narrowband" SNR, which is related to the coherent integration time, i.e., the time for which the successive samples of signals retain appreciable temporal covariance. The narrowband SNR can be obtained by using Eq. (3), with B equal to the inherent bandwidth of the lidar backscatter signal. This signal bandwidth is approximately equal to the reciprocal of twice the atmospheric coherence time measured at the 1/e point of the correlation function[14], assuming that the pulse coherence time is larger than the atmospheric coherence time. The narrowband SNR cannot be less than unity (0 db) due to the restriction that the photon flux must be not less than that required for one photo-mixing event per correlation time; otherwise, no velocity estimate is possible. In this Doppler lidar case, the independent temporal samples of return signal intensity (photon flux) are exponentially distributed due to the speckle effect. Thus an estimate may be made a fraction of the time, even if the average photon arrival rate is below that given by the above criterion. The resulting reduction in the number of valid velocity estimates within the range gate time may be calculated[4] using the exponential probability distribution of photon flux.

V. DOPPLER LIDAR MEASUREMENTS IN THE ATMOSPHERE

Initial applications of Doppler lidar for atmospheric measurements incorporated cw lidar systems. These systems are useful in studies where short range (less than about 200 m) are acceptable and where very small scattering volumes are desired. With the advent of pulsed coherent lidar technology, the range of applications of Doppler lidar has increased significantly. In 1981, the Wave Propagation Laboratory of the National Oceanic Atmospheric Administration (NOAA) began using the first pulsed TEA CO_2 laser system for the measurement of the atmospheric winds. The NOAA lidar included a hybrid TEA transmitter, heterodyne receiver, and digital signal processor. The lidar was entirely self-contained in a transportable semitrailer, and was deployed in a number of field studies. To show that coherent TEA laser transmitters with energies of 1 J and higher could be developed for space applications, NOAA upgraded the coherent lidar system transceiver in 1985, replacing the hybrid transmitter with an injection-locked TEA laser. The output pulse energy of about 1 J available from the upgraded transceiver is obtained by means of an unstable resonator configuration. The system is usually operated with a prf of 10-20 Hz for atmospheric studies.

Although a Doppler lidar system measures only radial velocity, three-dimensional winds profiles can be obtained by the velocity-azimuth display (VAD) method. With this technique, the lidar beam is positioned at a constant elevation angle and scanned at a fixed rate through 360 degrees in azimuth. When the wind flow is approximately laminar, and when there is no horizontal shear across the base of the cone circumscribed by the lidar beam, the plot of measured radial velocity versus azimuth for a lidar VAD scan produces a nearly sinusoidal wave with fluctuations due to measurement noise and small-scale turbulence present in each radial velocity measurement. By fitting a sine wave to the observed data by means of a least-squares technique, the wind field characteristics can be estimated: amplitude of the sine wave is proportional to the average horizontal wind speed; phase is proportional to wind direction, and d.c. offset is proportional to vertical wind speed and/or horizontal divergence. The NOAA lidar can also be raster scanned over a specified range of azimuth and elevation in order to study flow patterns which may be confined by local terrain. This capability was used[15] to study nocturnal drainage winds in a Western Colorado valley. Studies of downslope windstorm events and mountain waves near Boulder, Colorado have also been conducted when using the lidar either in a fixed pointing configuration, observing features which were advected through the beam, or in a raster scan configuration.

Although Doppler lidar as a remote sensor for atmospheric observations is still in its infancy, it is clear from the examples obtained with the NOAA lidar that the technique can fill data voids that are beyond the capabilities of other, more traditional atmospheric sensors.

VI. DOPPLER LIDAR FOR GLOBAL WIND FIELD MEASUREMENTS

In the late 1970's several institutions organized studies and workshops to discuss the feasibility of global atmospheric wind field measurements from Earth orbit. These included NASA Marshall Space Flight Center, NOAA WPL, and JPL. These feasibility studies resulted in an instrument concept which has served as a baseline for many further studies of various aspects of a Doppler lidar global wind measurement instrument. The instrument concept included CO_2 lidar technology, largely because of its clearcut technological maturity relative to other candidates. Since that time alternate lidar techniques have been proposed, some using direct detection instead of coherent detection. A comparative analysis of the various candidate Doppler lidars which was first presented at the NASA Global Winds Workshop and later published[4] in the open literature pointed out that the coherent detection lidars in the infrared had an advantage in overall photon efficiency, and that the CO_2 lidar operating in the 9 μm wavelength region had the additional advantages of eye safety, less stringent optical and pointing tolerances, and maturity when compared with the Nd-YAG coherent lidar, the only other technologically feasible coherent lidar candidate in this era.

NASA has recently become convinced that a Doppler lidar, given the name LAWS (Laser Atmospheric Wind Sounder), should become a major part of the Earth Observing System (EOS) and be placed on a polar orbiting platform in the late 1990's. According to the present concept, the instrument would contain a CO_2 laser transmitter emitting pulses of 10 J energy, at a prf of at least 10 Hz. A transmit/receiver mirror of approximately 1 m diameter would collimate the outgoing pulsed beam and collect and focus the backscattered radiation from the atmosphere. Conical scanning would be employed in order to achieve a swath width of approximately 2000 km at the Earth's surface, enough for nearly complete global coverage twice daily. Because of the alignment tolerances of the coherent receiver, lag-angle compensation optics would be necessary. Horizontal resolution would be 100-300 km and vertical resolution would be approximately 1 km from the surface to the top of the troposphere.

The baseline CO_2 coherent lidar instrument concept is providing many technological challenges; however, as important as the technology development activities have been, the performance criteria for the orbiting Doppler lidar have hinged on estimates of global tropospheric aerosol backscatter coefficients. In early 1986 NASA formed a panel directed by the Marshall Space Flight Center to plan a Global Backscatter Experiment (GLOBE), which consists of assessing the best available data sources, supporting and organizing on-going aerosol backscatter measurement programs, and planning aircraft flights to add to the knowledge of the global distribution of aerosol backscatter at IR wavelengths, the most important being the 9 μm wavelength of the CO_2 lidar. To date, the chief sources of aerosol backscatter measurements at infrared wavelengths in the troposhere and lower stratosphere are the CO_2 lidar backscatter profile data from Boulder, Colorado (NOAA WPL)[16] over the 1981-1984 period and from Pasadena, California (JPL)[17] over the 1984-present time period, along with flight data using the airborne LATAS (focused cw CO_2 Doppler sensor) instrument of the Royal Signals and Radar Establishment[18]. In addition, aerosol physical data from a number of field measurement programs have been combined with Mie scattering computations to estimate backscatter coefficients in the 10 μm region, and SAGE extinction data at 1 μm wavelength have been analyzed extensively[19] in order to infer a global climatology of infrared aerosol backscatter in the upper troposphere and lower stratosphere. The additional information gained from extended flights over the Pacific Ocean basin, including the southern hemisphere, should result in modelling atmospheric backscatter properties for LAWS sensitivity studies with a much greater level of accuracy.

VII. CONCLUSIONS

Increasing recognition is being given to Doppler lidar as a tool in the study of atmospheric dynamics. The capabilities of coherent Doppler lidar have been and vill continue to be driven by advances in laser, optical, photodetector, and signal processing technologies. However, the present level of maturity is sufficient for a variety of intensive field studies. The potential of a LAWS class instrument in Earth orbit, generating a unique set of wind field data on a global scale provides promise of a bright future and a motivation for advances in the field during the next several years.

REFERENCES

1. D. Rees, P. A. Rounce, I. McWhirter, A. F. D. Scott, A. H. Greenaway and W. Towlson, J. Phys. E:Sci.Instrum., 15, 191-206 (1982)
2. V. J. Abreu, Appl. Opt., 18, 2992 (1979)
3. I. S. McDermid, J. B. Laudenslager and D. Rees, UV-Excimer laser based incoherent Doppler lidar system, in: "Global Wind Measurements", W. E. Baker and R. J. Curran, Eds. (A. Deepak, Hampton, VA (1985)
4. R. T. Menzies, Appl. Opt., 25, 2546-2553 (1986)
5. R. T. Menzies and R. M. Hardesty, Proc. IEEE, 77, 449-462 (1989)
6. A. E. Siegman, Proc. IEEE, 54, 1350-1356 (1966)
7. R. H. Kingston, "Detection of Optical and Infrared Radiation", Springer-Verlag, Berlin/Heidelberg/New York, 1978, ch. 3
8. B. J. Rye, Appl. Opt., 18, 1390-1398 (1979)
9. J. H. Shapiro, Appl. Opt., 26, 3600-3606 (1987)
10. A. E. Siegman, Appl. Opt., 13, 353-367 (1974)
11. A. Parent, N. McCarthy and P. Lavigne, IEEE J. Quant. Electron., QE-23, 222-228 (1987)
12. D. M. Tratt and R. T. Menzies, Appl. Opt., 27, 3645-3649 (1988)
13. R. J. Doviak and D. S. Zrnic, "Doppler Radar and Weather Observations", Academic Press, New York (1984)

14. R. J. Keeler, R. J. Serafin, R. L. Schwiesow and D. H. Lenschow, J. Atmos. and Oceanic Tech., 4, 113-127 (1987)

15. M. J. Post and W. D. Neff, Bull. Amer. Meteorolog. Soc., 67, 274-281, (1985)

16. M. J. Post, Appl. Opt., 23, 2507-2509 (1984)

17. R. T. Menzies, G. M. Ancellet, D. M. Tratt, M. G. Wurtele, J. C. Wright and W. Pi, J. Geophys. Res., 94, 9897-9908 (1989)

18. J. M. Vaughan, D. W. Brown, P. H. Davies, C. Nash, G. Kent and M. P. McCormick, Nature, 332, 709-711 (1988)

19. G. S. Kent and S. K. Schaffner, Analysis of atmospheric dynamics and radiative properties for understanding weather and climate; Task 1. 10 μm backscatter modeling, "STC Technical Report 2175", (Science and Technology Corp., Hampton, Virginia, January 1988)

DIFFERENTIAL OPTICAL ABSORPTION SPECTROSCOPY (DOAS) FOR THE CONTINUOUS

MONITORING OF ATMOSPHERIC TRACE CONSTITUENTS AT A HIGH ALPINE STATION

B. E. Lehmann

Physics Institute
University of Bern
Bern, Switzerland

I. INTRODUCTION

Conventional trace gas analyzers for pollution monitoring generally operate as gas samplers at a fixed site. In areas with large spatial variation it is therefore not always easy to determine how representative the measurements are. Furthermore, each trace substance such as SO_2, NO_2 or O_3 requires a separate instrument, and these instruments may not be sensitive enough to determine low concentrations under clean air conditions. For example, a typical detection sensitivity for SO_2 or NO_2 is in the range of 1-2 ppb.

With the development of the DOAS-technique[1,2], especially when combined with adequate computer based data acquisition, a flexible and highly sensitive instrument for the simultaneous measurement of a variety of trace substances has become available. Optical absorption measurements over a path length of up to 10 km through the atmosphere enable monitoring of NO_2, SO_2, H_2CO and other pollutants in the sub-ppb range and at the same time the monitoring of other species (e.g., O_3 and H_2O) at their ambient levels.

Under the auspices of the European Program COST 611 "Physical and Chemical Behaviour of Atmospheric Pollutants", a commercial DOAS system has been installed at the Swiss High Alpine Research Station at the Jungfraujoch in the central Swiss Alps (3450 m a.s.l., elevation above sea level). It will contribute to a better understanding of the transport of polluted air masses over a complex topography.

II. TECHNICAL DATA ON THE ANALYTICAL SYSTEM

The system is commercially available (OPSIS, Sweden). It was described in its essential components by Edner[3] et al., 1986. An overview of the instrument is presented in Figure 1. The light source is a high pressure

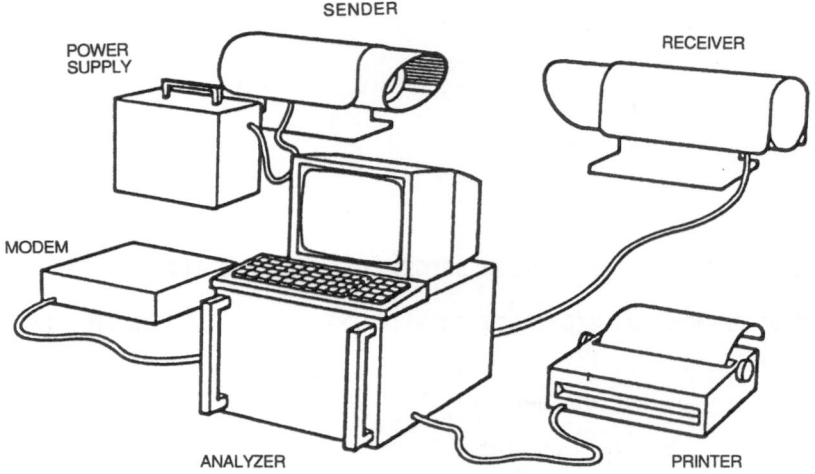

Fig. 1. Components of the commercial DOAS system (OPSIS).

150 Watt Xenon lamp placed in front of a 30 cm diameter concave mirror with a focal length of 25 cm. It produces a well collimated beam of about 1 mrad divergence. This part of the instrument together with the high-voltage power supply has been mounted at the Jungfrau-Ostgrat (3700 m a.s.l.). It remains outdoors during operation and therefore had to be placed on a solid mount to guaranteee optical alignment even under bad weather conditions. An extra heater prevents the formation of ice on the front window and a steep roof-like cover protects the source from snow accumulation. The system has now continuously been operated for eight months with the original xenon lamp. To our surprise no adjustments at all were necessary during this time.

After a 968 meter free path through the atmosphere the light is collected by the receiver, which is placed in a laboratory at Research Station Jungfraujoch at 3450 m. It is focussed by a second spherical mirror of 15 cm diameter and 90 cm focal length onto a optical fiber, by which it is guided to the entrance slit of a fast scanning spectrometer.

For each trace substance to be analyzed the grating is set to allow a scanning over a range of about 40 nm centered in the corresponding wave-length region where the particular molecule absorbs light. The scanning is accomplished by a fast rotating disk with optical slits. One hundred spectra are recorded per second and the photomultiplier signal is diverted by a fast A/D converter into 1000 channels. One channel therefore repre-sents a spectral width of about 0.04 nm. With an integration time of e.g. 5 minutes a total of 30,000 spectra is accumulated before a concentration value is calculated. For this calculation the data acquisition program compares the spectrum with calibration spectra of the molecule under investigation. Then the system switches to the next substance for anal-ysis.

Table 1. Central wavelengths for five molecules
and detection limits for a 2 minute
integration time.

Component	Central Wavelength	Detection limit
O_3	270 nm	3 ppb(V)
SO_2	300 nm	0.1 ppb(V)
HNO_2	350 nm	0.1 ppb(V)
NO_2	430 nm	0.2 ppb(V)
H_2O	720 nm	0.2 g/cm^3

Table 1 gives the central wavelength for each component and the
detection limit for an integration time of 2 minutes. The OPSIS system
is connected through a modem to the public telephone network and therefore
on-line system checks are always possible from any PC at home (for
instance at night when the outdoor temperature at the Jungfraujoch is
below - 30 °C and strong winds shake the emitter). One necessary condition
for a measurement of course is that the visibility is better than 968 m.

III. FIRST RESULTS

Figure 2 illustrates the kind of results one can expect with such a
system. It represents a period of nine days from January 19 to January 27,
1989 and gives an interesting sequence of various air masses that have
reached this high alpine environment[4]. Peaks of SO_2 that extend over
several hours up to one day are clearly visible. During the first SO_2 peak
a slight increase in NO_2 can be recognized while the last SO_2 peak coincides with a minimum in NO_2. The second SO_2- contaminated air obviously
was transported by a cold and humid air mass (lowest two records) and it
is interesting to note than an increase in O_3 occured prior to the SO_2
increase.

IV. SUMMARY

A commercial DOAS system has been operated for a period of eight
months. It appears to be well suited for continuous high-sensitivity
measurements even under extreme environmental conditions. The detection
sensitivity for SO_2 is low enough to clearly recognize episodes of
increased pollution that occur even at very remote sites. NO_2 variations
generally appear to be lower, and it is planned to further improve the
detection sensitivity for this gas by changing to a system with a doubled
path length using a retroflector at the site where the light source is
currently installed. The alternating monitoring of SO_2, NO_2, O_3 and H_2O in

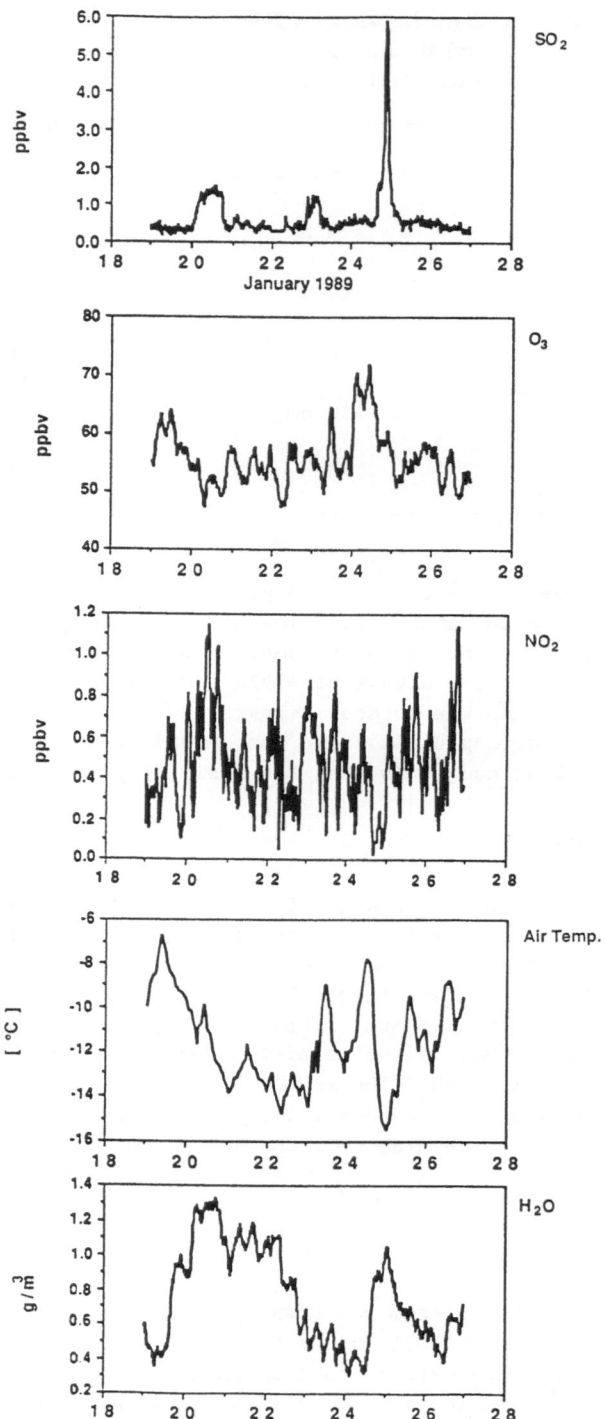

Fig. 2. DOAS data for the period 18 to 27 January 1989. From top to bottom: SO_2 (0 to 6 ppb); NO_2 (0 to 1.2 ppb); O_3 (40 to 80 ppb); Temp. (-16 °C to -6 °C); and, H_2O (0.2 to 1.4 g/m³).

120

sequences of 5 minutes, together with a continuous monitoring of the relevant meteorological parameters (pressure, temperature, wind, solar irradiation) and with simultaneous measurements of peroxides, of aerosols and of radon, allows a very distinct characterization of air masses that are being transported to this high alpine site. Such data will be combined with those from other stations and used to calibrate mesoscale transport models.

V. ACKNOWLEDGEMENTS

The following scientists at the Physics Institute in Bern have contributed to this work: A. Neftel, M. S. Lehmann. Technical assistance from H.-P. Moret, H. Riesen and K. Grossenbacher is greatly appreciated.

REFERENCES

1. U. Platt, D. Perner and H. W. Pätz, J. Geophys. Res., 84, C10, 6329 (1979)
2. U. Platt and D. Perner, J. Geophys. Res., 85, C12, 7453 (1980)
3. H. Edner, A. Sunesson, S. Svanberg, L. Uneus and S. Wallin, Appl. Opt., 25, 403 (1986)
4. A. Neftel, B. E. Lehmann and S. Lehmann, Monitoring of Peroxides and Trace Gas Constituents at a high Alpine Station, Proceedings of the COST 611 Conference on the "Physical and Chemical Behaviour of Atmospheric Pollutants", Varese, Sept. 25-29 (1989)

PHOTOACOUSTIC NH_3 MONITORING WITH WAVEGUIDE CO_2 LASERS

M. Hammerich, A. Ölafson and J. Henningen

J. C. Orsted Institute
Physics Laboratory
University of Copenhagen
Köbenhavn, Denmark

Ammonia is not a natural constituent of fossil fueled power plant emission, but nitrogen oxides and sulphur dioxide are. In various schemes for denitrification and desulphurization, ammonia is introduced either as a reducer of NO_x as in the Selective Catalytic Reduction (SCR) process or as a neutralizer for acids developed in the denitrification, as in the corona based CORONOX process. In either case one wants to monitor the efficiency of the process and to be able to document compliance with NH_3 emission limits. Thus, there is an increasing need for ammonia measurements in smoke stacks, and the monitor needs to be:
- moderately sensitive, corresponding to 1 ppm detection limit;
- very selective, so as to monitor NH_3 regardless of known and unknown components in the smoke;
- automatic;
- moderately fast, corresponding to 1 measurement per 10 minutes for normal monitoring, and 1 per minute for reactor characterization and process optimization.

In order to meet these needs, we have built a computerized photo-acoustic laser spectrometer and field tested it at a Danish power plant.

In photoacoustic spectroscopy one shines a modulated beam of light through the sample and measures the acoustic field generated when excited molecules relax through collisions, transferring the absorbed energy to heat, and thereby to pressure variations in the gas at the modulation frequency. For laser application of this technique see Refs. 2, 3 and 4.

Referring to Figure 1, we will briefly discuss the main components of our system. The lasers we use are short, flowing gas, waveguide CO_2 lasers operating around 160 mbar, with pulsed excitation at a repetition rate of 700-900 Hz. The lasers can be tuned in windows of ± 230 MHz through 80 lines of the CO_2 laser spectrum. The laser line is selected by rotation

Optoelectronics for Environmental Science, Edited by S. Martellucci and
A.N. Chester, Plenum Press, New York, 1990

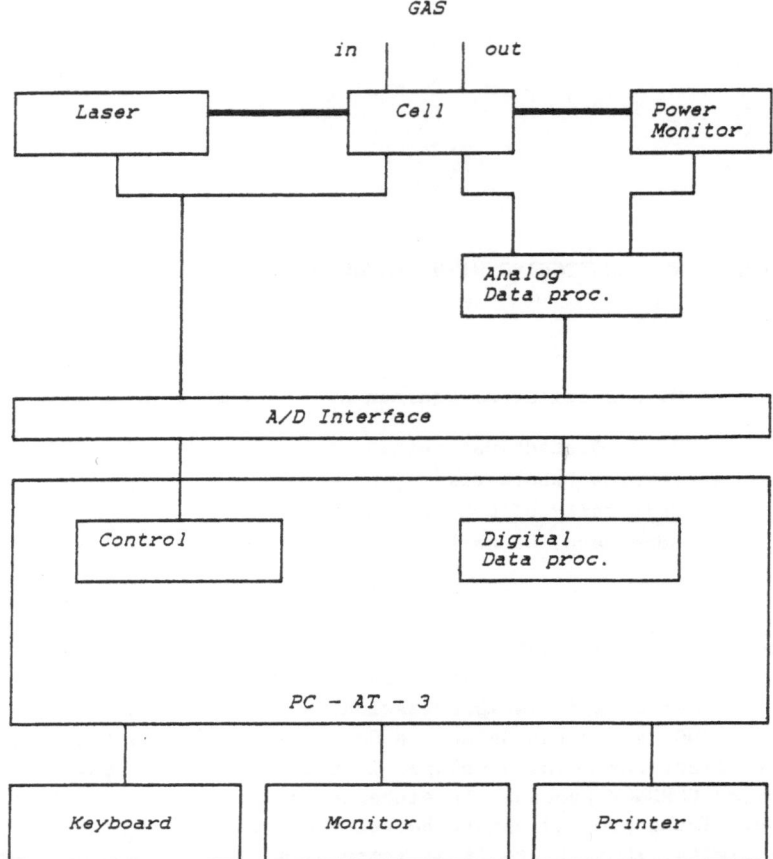

Fig. 1. Block diagram of the photoacoustic laser spectrometer.
From Ref. 1.

of a blazed grating (see Figure 2). It may be tuned over three free
spectral ranges by varying the length of the resonator with a piezoelec-
tric transducer.

To ensure single line and single mode tunability over the full free
spectral range of the laser, we fine tune the distance between the waveguide and the grating. When varying this distance to optimize the laser,

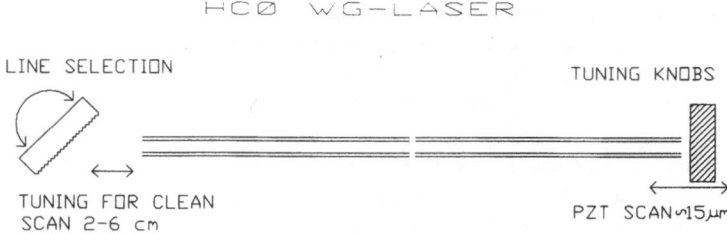

Fig. 2. Schematics of the CO_2 waveguide laser.

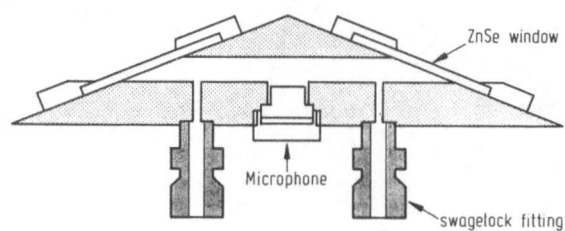

Fig. 3. Nonresonant photoacoustic cell. From Ref. 1.

two conditions must be simultaneously fulfilled. First, the distance must
be such that competing neighbouring lines tune away from their line
centers as the main line is tuned to the edges. If this were not so, there
would be parts of the scan where a neighbouring line sees a rising gain
while the gain in the main line goes down, and this very often leads to
line hopping. Second, a phase condition must be met such that the various
free space laser modes excited in the space between the waveguide and the
grating add up to a narrow field configuration upon reentrance into the
waveguide. For a more thorough discussion of this effect see Ref. 6.

The photoacoustic cell, shown in Figure 3, is a 10 cm by 8 mm
diameter nonresonant cavity, constructed from stainless steel. It is
equipped with Brewster windows of ZnSe and a 1/2" Bruel & Kjaer electret
microphone. The entire cell assembly and the gas handling system can be
heated to 120 °C in order to match the stack conditions and to minimize NH_3
adsorption. The sample flows through the cell at a rate of 50 ml/min STP.

In order to utilize the large tunability of our CO_2 laser to increase
the selectivity of the spectrometer, we operate the cell at a reduced
pressure of 12 mbar. Under these circumstances the rotation-vibration
lines of small molecules have mixed braodening with a line width of around
70 MHz. This offer two advantages: first, we are able to resolve a single
line of ammonia, which gives us a virtually unique fingerprint; second, we
suppress the contribution of other molecules by a large factor[1]. This is
extremely important when using CO_2 lasers on samples with a high content
of CO_2, since the laser transition will absorb, and give rise to a photoacoustic signal at line center. We detect ammonia at the sR(5,0) transition of the rotation-vibration spectrum. This line is situated -190 MHz
from the line center of the 9R30 CO_2 laser line, and we thus have ample
offset to discriminate between the two lines.

The data processing system is rather straigthforward. The amplitude
and phase of the photoacoustic signal are detected at the laser pulse
frequency by a dual lock-in amplifier. The laser power is used for normalizing the photoacoustic signal and is further used as basis for a

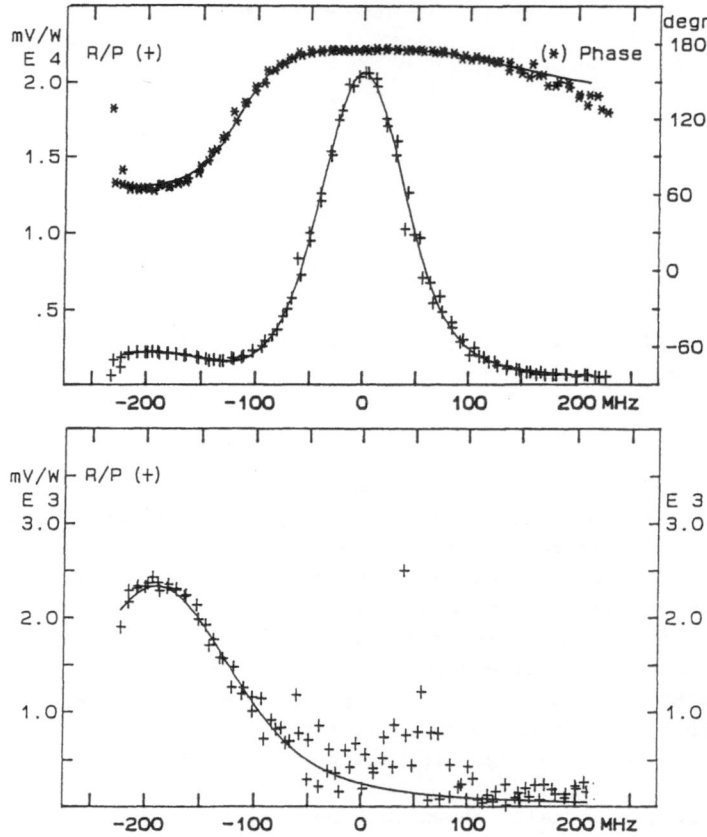

Fig. 4. Example of photoacoustic data taken at the field test at a Danish power plant. The sample consists of 11.7% CO_2, 9.2 ppm NH_3, some 8% H_2O and unknown amounts of NO_x and SO_2. In the lower frame the CO_2 signal has been subtracted. The cell was held at 100 °C and the pressure was 12 mbar. From Ref. 1.

linearize-cut-paste procedure for transforming the piezo scan of three free spectral ranges into a linear frequency axis. The photoacoustic signal is then fitted to a sum of two Voigt lines and a constant background, all with complex amplitudes. The magnitudes of the two Voigt lines are taken as a measure of the concentrations of the sample components.

In Figure 4, an example of data taken in a field test is shown. In the upper frame, the phase and magnitude of a signal from 11.7% CO_2 and 9.2 ppm ammonia is shown. The fully drawn curves represent the fit to the data. The contribution from NH_3 is seen as part of an usual line profile, and in the phase data as a significant change of phase. The reason for the phase contrast between the two signals lies in the molecular dynamics involved in the relaxation process. Adsorbed energy in the CO_2 molecules is resonantly transferred to the first metastable vibrational state in N_2

by collisions. The CO_2 vibrational ladder is now out of thermal equilibrium and vibrational energy is transferred to the lower level of the absorbing transition at the expense of translational energy. This process is known as kinetic cooling[7,8] since, instead of an immediate heating of the gas, we see a transient cooling. The water content of the gas speeds up the relaxation of N_2 molecules, so that the phase contrast is somewhat reduced. We use this phase contrast as a very important diagnostic tool in order to reveal any malfunction of the laser, since line or mode jumping will show up as a CO_2 signal located away from zero offset.

In the lower frame of Figure 4 the fit to the CO_2 line is subtracted from the data. The two lobes of the signal at -100 and +100 MHz are residuals from the CO_2 signal revealing that the frequency axis has not been perfectly established. However evaluation of the fluctuations between -235 and -100 MHz gives a detection limit for NH_3 of less than 0.5 ppm.

In conclusion, the results of the field test show that a tunable CO_2 laser can be used for ammonia measurements in industrial environments down to the 1 ppm level. The main advantage of spectrally resolved spectroscopy is that the ammonia concentration is derived from a spectral profile, and not just by comparison of two numbers, referring to strong and weak absorption respectively. This, in addition to the utilization of the photoacoustic phase, makes the system virtually immune to interference.

ACKNOWLEDGEMENTS

The authors wish to acknowledge the support of the Danish Science Research Council under grants no. 5.17.4.6.19 and 5.17.4.1.23 and the support and assistance of the Danish Power Utility Companies (ELSAM).

REFERENCES

1. A. Olaffson, M. Hammerich, J. Bulow and J. Henningsen, Appl. Phys., B 49, 91-97 (1989)
2. E. L. Kerr and J. Atwood, Appl. Opt., 7, 915-921 (1968)
3. "Photoacoustic and Photothermal Phenomena", P. Hess and J. Pelzl, eds., Springer Ser. Opt. Sci. Vol. 58, Springer, Berlin Heidelberg (1987)
4. V. P. Zharov and V. S. Letokhov, "Laser Optoacoustic Spectroscopy", Springer Verlag, Heidelberg (1986)
5. F. Tang and J. O. Henningen, Appl. Phys., B 44, 93-98 (1987)
6. R. Gerlach, D. Wei and N. M. Amer, IEEE Journ. of Quantum Electron., 20, 948-963 (1984)
7. F. G. Gebhardt and D. C. Smith, Appl. Phys. Lett., 20, 129 (1972)
8. A. D. Wood, M. Camac and E. T. Gerry, Appl. Opt., 10, 1877-1884 (1971)

OCEANOGRAPHIC MEASUREMENTS

FLUORESCENCE LIDARS IN ENVIRONMENTAL REMOTE SENSING

L. Pantani and G. Cecchi

C.N.R.
Istituto di Ricerca sulle Onde Elettromagnetiche
Firenze, Italy

I. INTRODUCTION

The word LIDAR is the acronym of LIght Detection And Ranging and appeared the first time in 1953[1]. The history of Lidars began with the invention of the giant pulse laser in 1962[2], which allowed its development. The first applications to the remote sensing of the atmosphere were published in 1963[3,4], while the first experiments on remote sensing of natural waters were carried out about ten years after[5-7]. Only in the last decade the use of lidar systems for the remote sensing of ground surface and vegetation was introduced[8-11].

The block diagram of a lidar system is shown on Figure 1. A collimated beam impinges on the target and a telescope collects part of the diffused radiation. The collected radiation is spectrally analyzed, and detected. A central computer controls the system, stores and pre-process the data.

The lidar analysis of the environment[12,13] is mainly based on two interaction processes between the laser light and the target: elastic scattering, where the detected wavelength λ_i is the same as the laser wavelength λ_0, and the inelastic scattering where $\lambda_i \neq \lambda_0$. Without multiple scattering[14] the power p diffused from the target at the wavelength λ_i which is received at time t_0 is given by:

$$p(R,t_0,\lambda_0,\lambda_i) = p_0(t_0-2R/c)(k/R^2)\eta(R)\beta(R,\lambda_0,\lambda_i)$$

$$\cdot \exp \{-_0\!\int^R[\alpha_0(r) - \alpha_i(r)] \, dr\} \tag{1}$$

where $p_0(t)$ is the laser power at the time t, R the target distance, K a constant and $\eta(R)$ a "geometrical form factor"[15] both depending on the system structure, $\beta(R,\lambda_0,\lambda_i)$ is a coefficient which takes into account the efficiency of the interaction process and the number of

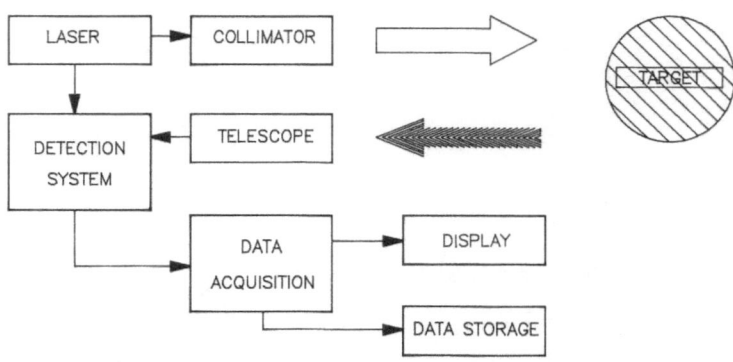

Fig. 1. Lidar principle.

molecules excited by the laser beam, and $\alpha_n(r)$ is the global extinction coefficient in the point r of the propagation medium at the wavelength λ_n.

When a pulsed laser is employed, the emitted power is greater than 0 only for the time T corresponding to the pulse duration. So the equation (1) can be integrated on t_0, in order to express the energy received, by scattering on a target at distance R, at a wavelength λ_i:

$$E(R,\lambda_0,\lambda_i) = (KE_0/R^2)\eta(R)\beta(R,\lambda_0,\lambda_1)\exp\{-\int_0^R[\alpha_0(r) - \alpha_i(r)]dr\} \quad (2)$$

where E_0 is the laser pulse energy. Equation (2) is one of the forms of the "lidar equation".

II. LIDAR SPECTROSCOPY[12,13]

The interaction processes between the laser radiation and the target itself are at the basis of the information which can be extracted from a Lidar signal. The equation (2) contains two terms which express this kind of interaction. The first, $\beta(R,\lambda_0,\lambda_i)$, indicates the part diffused by the target in the direction of the receiver, while the second, $\alpha_n(r)$, indicates the radiation losses due to absorption and scattering on different direction from that of the receiver.

The spectrum of the backscattered radiation gives information on target characteristics. Scattering phenomena can be described in two main processes: elastic scattering and inelastic scattering (Figs. 2 and 3).

In the elastic scattering the received wavelength is the same of the excitation one ($\lambda_0 = \lambda_i$); two cases are possible: Mie scattering and Rayleigh scattering. Mie scattering (Fig. 2a) is related to the classic elastic scattering which takes place when the dimensions of the particle are close to or larger than the optical wavelength. Rayleigh scattering is known to occur in phase with the incident radiation, without any appreciable energy exchange with internal states of atom or molecule, as depicted in Figure 2b. For this reason the elastically scattered light does

132

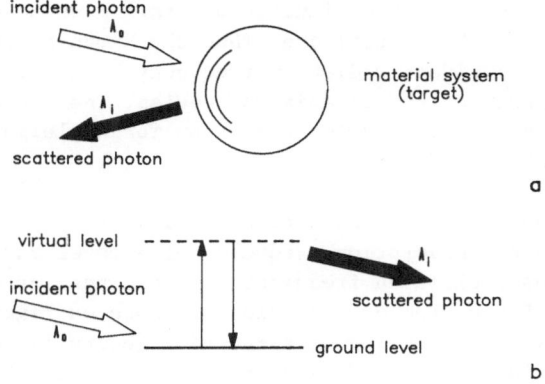

Fig. 2. Elastic scattering: (a) Mie scattering;
(b) Rayleigh scattering.

not identify the scatterer (in this case the target), but only monitors the geometrical characteristics of the target.

In the inelastic scattering the monitored wavelength is different from the excitation one ($\lambda_i \neq \lambda_0$); in this case an appreciable energy exchange between the incident radiation and the internal states of target material. As a consequence, the inelastic scattering gives information on chemical-physical parameters of the target. In Lidar remote sensing of sea and vegetation a particular interest is devoted to Raman scattering and fluorescence. Raman scattering is the process schematized in Figure 3a, concerning the excitation of a virtual level (and therefore with a zero

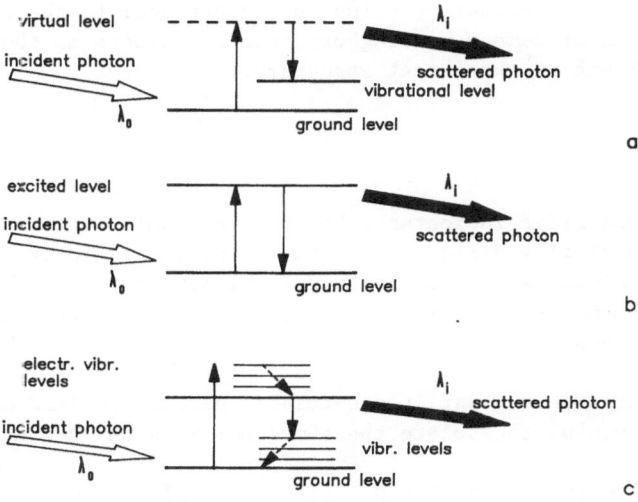

Fig. 3. Inelastic scattering: (a) Raman scattering;
(b) Resonant Raman scattering; (c) Fluorescence.

lifetime) of the target and the simultaneous fall down to a real level, but anyhow different from the initial one. Thus the Raman scattering component is shifted from the incident radiation frequency by an amount corresponding to the internal energy of the irradiated species. The cross section for Raman scattering is usually smaller than that for Rayleigh scattering by about three orders of magnitude.

Since Raman scattering cross-section is inversely proportional to the difference between the absorption frequency of a level and the excitation frequency, when the excitation frequency is tuned near or close to a proper resonance of the atom or molecule, as shown on Figure 3b, a great increase in the Raman scattering cross-section generally takes place. This process is conventionally called resonance Raman scattering, and includes not only exactly on-resonance but also near-resonance interactions. Fluorescence is the spontaneous emission (Fig. 3c) of a photon following excitation into an excited state by absorption of incident radiation within a specific absorption line or band of an atomic or molecular species. The excited level decays by re-emitting photon via transitions to the original and different lower levels. The excited atoms and molecules also suffer collisions which redistribute them into other excited levels through nonradiative transitions, as denoted by the dashed arrows in Figure 3c. Fluorescence is generally thought of as two single-photon processes; that is, a two-step interaction consisting of absorption of a single photon of a proper frequency followed by spontaneous emission of a photon with different frequency. These two steps involve an intrinsic time, characteristic of the system, which is called fluorescence lifetime. The fluorescence lifetime is of the order of a nanosecond in complex organic molecules or lower, and can be utilized in target identification, even if sophisticated excitation and detection system are required.

Both fluorescence and Raman scattering contain elements (that is, emission frequency and lifetime in a case, and frequency shift in the other one) which are characteristic of the scatterer. The signal given by both processes has an intensity which is proportional to the scatterer density. By means of remote sensing of these parameters is then possible to have an analysis of the target structure.

III. FLUORESCENCE LIDAR

Fluorescence Lidar indicates a Lidar system which detects the radiation unelastically diffused from the target. This feature means that a fluorescence Lidar can be, at least potentially, a "Raman" Lidar too, and, indeed, Raman signal is largely used in natural water remote sensing, how it will be shown later on.

Since fluorescence Lidar is essentially devoted to land and sea remote sensing, it is useful to isolate the atmospheric terms in Equation 2:

$$E(R,\lambda_0,\lambda_i) = (KE_0/R^2)n(R) \ \beta(R,\lambda_0,\lambda_i) \ \exp \ \{-H[\alpha_0(H)+\alpha_i(H)]+$$

$$-_H\int^R [\alpha_0(r) + \alpha_i(r)]dr\} \tag{3}$$

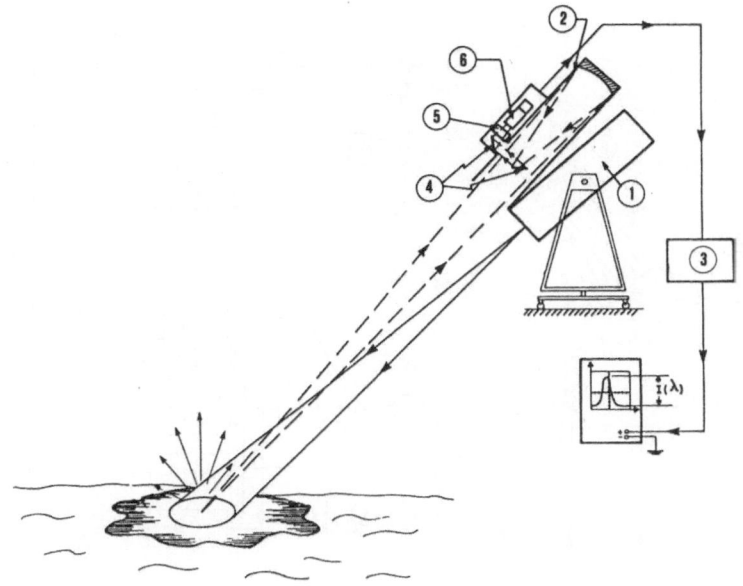

Fig. 4. Single channel lidar for oil remote sensing[16].
(1) N_2 laser, (2) Newtonian telescope, (3) signal
processing, (4) stearing mirrors, (5) interference
signal, (6) photomultiplier. The received signal
is monitored on an oscilloscope.

H being the atmospheric propagation path and $\alpha_n(H)$ the mean value of
$\alpha_n(r)$ on H.

Fluorescence Lidar had its first application in oil teledetection.
Since these substances have a strong absorption in the ultraviolet,
nitrogen laser (337.1 nm) previously and excimer laser (308 nm) subsequently were selected as excitation sources. Chlorophyll fluorescence,
which is detected for phytoplankton and vegetation monitoring, needs an
excitation in the visible and for this application dye laser, optically
pumped by lamps or ultraviolet laser, is used.

The receiving system has different channels, chosen on the needed
wavelengths. Channel selection is usually made by means of a grating
spectrometer, or of an interference filter set.

On Figure 4 a sketch of one of the first fluorescence Lidar is
shown[16]. The system uses a nitrogen laser, has one receiving channel, and
it has been utilized both from aircraft and terrestrial platform. A more
advanced system is described on Figure 5: it is the airborne fluorescence
Lidar of the Canadian Centre for Remote Sensing[17], which has a nitrogen
laser, 16 receiving channels and an extra channel, placed on the laser
wavelength and time resolved, for altimetric use. Data are automatically
corrected for laser energy, detector spectral response, and aircraft

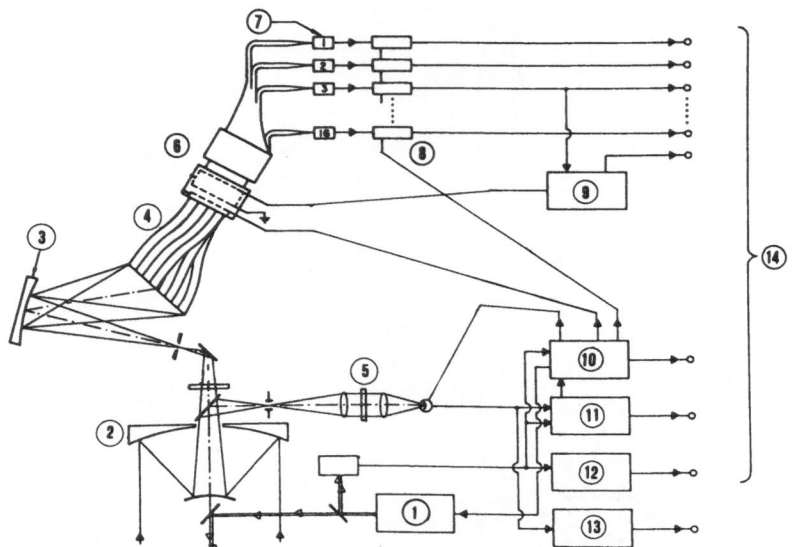

Fig. 5. Lidar of the Canadian Center for Remote Sensing[17].
(1) N$_2$ laser, (2) Cassegrain telescope, (3) concave
grating, (4) fiber optics coupling, (5) channel at
337.1 nm, (6) intensifier, (7) photodiodes,
(8) sample & hold, (9) automatic gain control,
(10) timing and gate, (11) Lidar altimeter,
(12) laser power measurement, (13) backscattered
signal amplitude, (14) to signal processing.

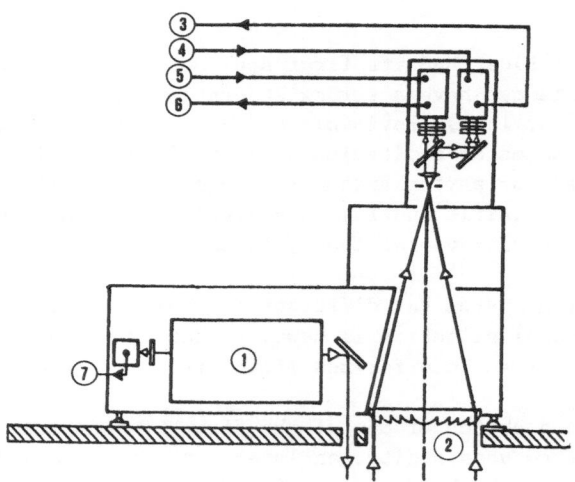

Fig. 6. Lidar for detection of chlorophyll $\underline{a}$[18]. Lamp
pumped dye laser, (2) Fresnel lens, (3) fluores-
cence signal, (4) and (5) gate, (6) Raman signal,
(7) laser power measurement.

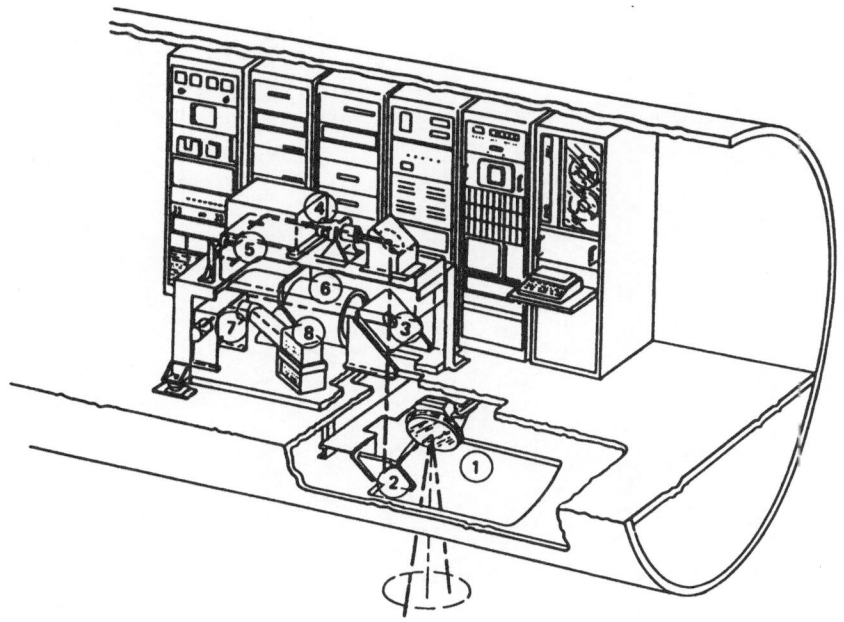

Fig. 7. The Lidar of NASA Wallops Flight Center installed inside
a C-54 carrier[19]. (1-3) Steering mirrors, (2) scanning
mirror, (4) collimator, (5) laser, (6) telescope, (7)
bathymetry detector, (8) fluorescence detector. At the
wall the data acquisition system.

height, while background radiation is measured by opening the gate in the
time interval between two laser shots and subsequentially subtracted.

A simple system for the aircraft remote sensing of phytoplankton[18] is
pictured on Figure 6. It uses a lamp pumped dye laser, two receiving channels, working on chlorophyll fluorescence band and on water Raman wavelength.

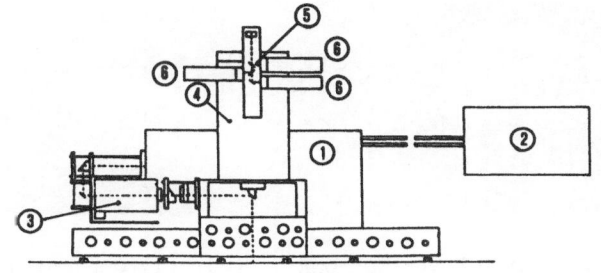

Fig. 8. Oldenburg University Lidar[21]. (1) Excimer
laser, (2) power supply, (3) dye laser, (4)
Schmidt telescope, (5) optical channel
splitters, (6) photomultipliers.

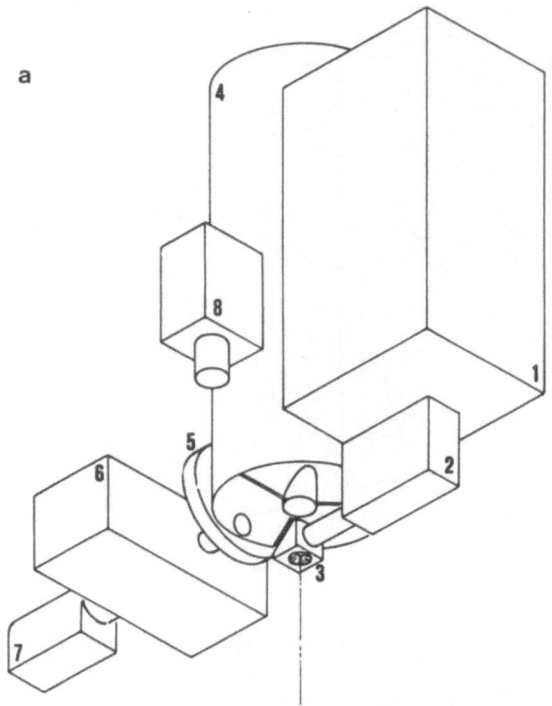

Fig. 9. The FLIDAR-2 of IROE-CNR[34]. (a): (1) excimer laser,
(2) dye laser, (3) steering mirror, (4) Newtonian
telescope, (5) filter wheel, (6) spectrometer, (7)
detector; and, (b): the FLIDAR-2 on board of the
Oceanographyc Ship Minerva (October 1987).

The most complex system has been realized at NASA/Wallops Island Flight Center (Fig. 7), utilizing a grating spectrometer with 40 PMT channels. The excitation wavelength is done with a nitrogen laser for oil detection[19] and with doubled Nd-Yag (532 nm) for bathymetry and phytoplankton monitoring[20]. This system has a conical scan for imaging. When excimer lasers became available, they practically replaced N_2 lasers in Lidar systems, because of their better characteristics.

The first fluorescence Lidar working with an excimer laser was developed at Oldenburg University[21] and it is shown on Figure 8. The laser uses a XeCl gas mixture which lases at 308 nm; the ultraviolet radiation is used both for oil detection and for pumping a dye laser, which gives a proper wavelength for the detection of gelbstoff and other dissolved matter in the water column. The receiving system has 6 channels, working with interferential filters and photomultipliers, the output signals are sent to a transient recorder, which allows both a spectral and spatial resolution, even if this is limited at six wavelengths.

The FLIDAR-2, developed and built at IROE-CNR in Florence (Italy)[22], is the most recent system and it is the first of a new generation of fluorescence Lidar for remote sensing of sea and vegetation. The systems is shown on Figure 9: it utilizes a XeCl laser[23] and a dye laser expressly designed for this application. The receiving system is composed of a grating spectrometer and an intensified gatable CCD array with 512 channels. The FLIDAR-2 has low weight and small size and it allows the contemporary detection of high resolution spectra for both reflectance and laser excited fluorescence and Raman emission.

In Table 1 the main characteristics of the above mentioned Lidars are reported.

Table 1. Some relevant examples of fluorescence lidars.

INSTITUTION	LASER	TELESCOPE	DETECTION	USE
Toronto University	N_2	unknown	PMT (1 chnl)	water pollution
NASA/WFC	N_2	30 cm	PMT (40 chnls)	water pollution vegetation
CCRS/Ottawa	N_2	20 cm	PD+Int (16 chnls)	water pollution
OLS/Oldenburg University	XeCl+Dye	40 cm	PMT (7 chnls)	water pollution vegetation
IROE-CNR/Firenze	XeCl+Dye	25 cm	CCD-Array+Int. gate (512 chnls)	water pollution vegetation

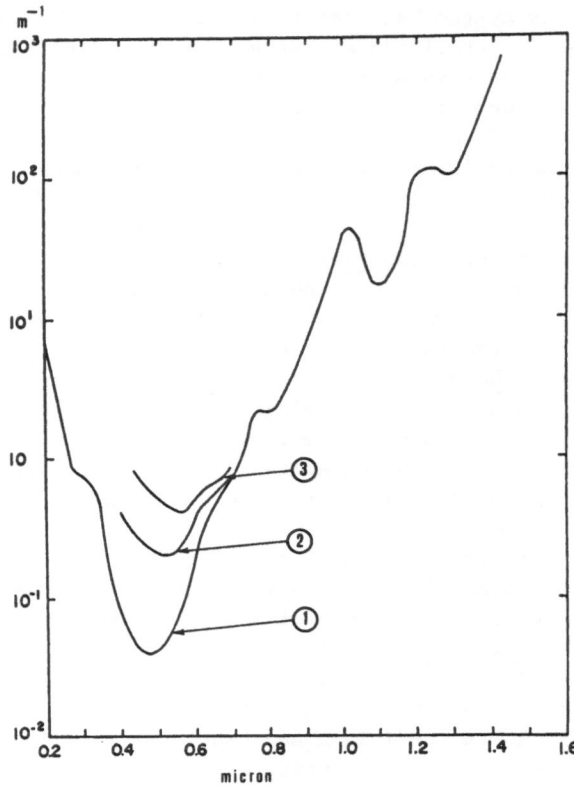

Fig. 10. Global extinction coefficient of water as a function
of wavelength[16]. (1) Distilled water, (2) coastal
waters, (3) estuarine waters.

IV. REMOTE SENSING OF NATURAL WATERS

Lidar systems are very well known in natural water remote sensing.
In fact the good penetration depth of a visible wavelength into water
(Fig. 10) and the possibility of a good range resolution make the Lidar
the only sensor capable to give informations on the water column with depth
resolution.

In the following the remote sensing of fluorescence and Raman radiation will be analyzed, omitting the application of elastic scattering, like
bathymetry.

IV.1. Water Turbidity

The Raman signal of OH stretching of water molecule shows a frequency
shift of 3400 cm^{-1} with respect to the excitation frequency. If the excitation is in the near ultraviolet, the Raman process is near resonant and
the cross section increases drastically (see the previous Section). Both
characteristics, that is near-resonance and large frequency shift, turn

useful in remote sensing, because they give a strong signal well-shifted from the excitation frequency, which is easily detectable. The strong intensity is moreover due to the high molecule concentration, fact which does not occur in atmospheric applications.

For what concerns the signal calculation, the concentration of water molecules can be assumed constant with R, so $\beta(R,\lambda_0,\lambda_i)$ in Equation (3) is constant. The signal is proportional to the term which depends on global extinction coefficient

$$E(R,\lambda_0,\lambda_i) \propto R^{-2}\exp\{-2\int_H^R \alpha_{0i}(r)dr\} \tag{4}$$

being $\alpha_{0i}(r)$ the mean value of the global extinction coefficient at the two wavelengths; with Equation (4) it is possible to extract the $\alpha_{0i}(r)$ behaviour and thus that of water turbidity. The integral over R of Equation (4) gives the global energy E_g, received from Raman emission:

$$E_g(\lambda_0,\lambda_i) = \int_H^\infty E(R,\lambda_0,\lambda_i)dr \propto (1/H^2)(1/\alpha_M) \tag{5}$$

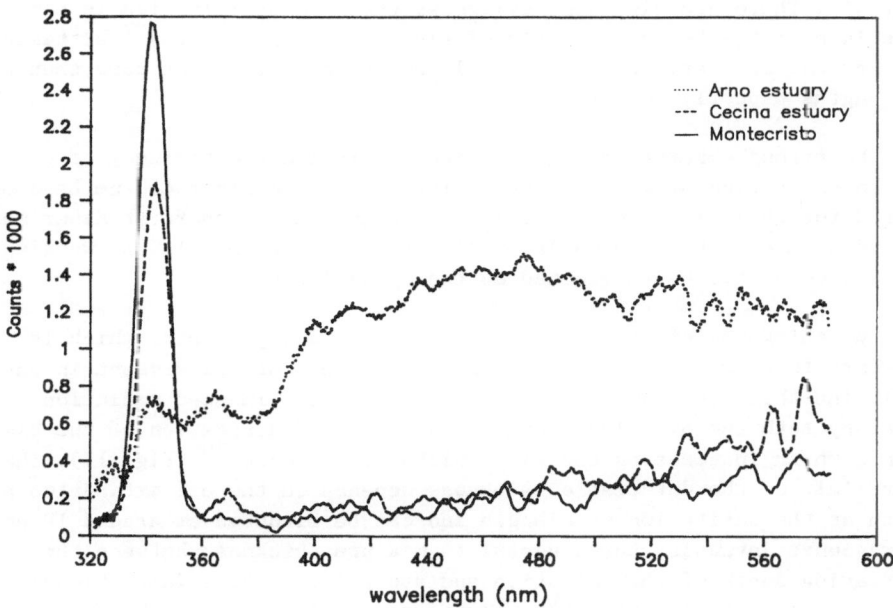

Fig. 11. Sea water turbidity, spectra detected with the FLIDAR-2 during the MARET experiment (October 1987). 1) Clear water (Montecristo island), 2) medium turbidity water (Cecina river front), 3) high turbidity water (Arno river front). Excitation wavelength 308 nm. Note at left the depression of the Raman peak when the turbidity increases.

α_M being the mean value of $\alpha_{0i}(r)$ along the propagation path into water. So the intensity of OH Raman stretching signal can be used as a measurement of the water turbidity or of the penetration depth (Fig. 11).

IV.2. Oil Remote Sensing

Due to the high quantum fluorescence efficiency of oils, oil remote sensing on sea water surface has been one of the first applications to sea remote sensing of fluorescence Lidar. Other parameters, like microwave and Thermal InfraRed (TIR) emission, allow the detection of oil spills over the water surface[24]. Anyway the oil spill are detected through some effects, changes of the surface apparent temperature or of the wave spectrum, which can be caused also by other phenomena, like wind slicks and surface currents, and show therefore a quite high false alarm probability. Since fluorescence and Raman techniques have a very low false alarm probability and show an unique potential in film thickness measurements, oil identification, detection of underwater and infra-ice oil, fluorescence Lidars can be at least an important complementary sensor in oil spill remote sensing.

The analysis of both fluorescence spectrum and time decay allows the identification of the oil which composes the spill. It was demonstrated[25] that it is possible to identify the oil class (crude, light, heavy) with correlation operations on the fluorescence spectrum or by the analysis of the ratio between the fluorescence emission at two properly selected wavelengths[26]. There are also some evidences that a discrimination inside the class is possible by the analysis of fluorescence time decay[27] or taking into account the ratio between the fluorescence emission at more than two wavelengths properly selected.

The strong absorption and the high fluorescence efficiency of oils with an excitation between the near ultraviolet and about 450 nm have been studied for the selection of the laser source. While the first experiments have been made with nitrogen laser[6,17], then the excimer laser, working at 308 nm (XeCl), has been selected as the optical choice[28].

The thickness of the oil film is an important parameter which is necessary to know in order to evaluate the amount of oil present in the spill. The thickness can be deduced from the laser induced radiation signal by both the oil fluorescence yield and the depression of the Raman signal, which increase as the film thickness increases[29] (Fig. 12). The upper limit of the detectable thickness depends on the oil extinction path length at the ecxitation wavelength and can be situated at around 10 μm. Fluorescence intensity turns useful to measure thickness between the penetration depth of the radiation and about 1 μm, while Raman Signal turns useful for lower thicknesses[29]; thicknesses as low as 10 nm have been detected[28]; the Lidar system is therefore an attractive sensor for the detection of very thin biogenic films.

IV.3. Water Pollution Detection

The detection and analysis of oil spills discussed above is an aspect

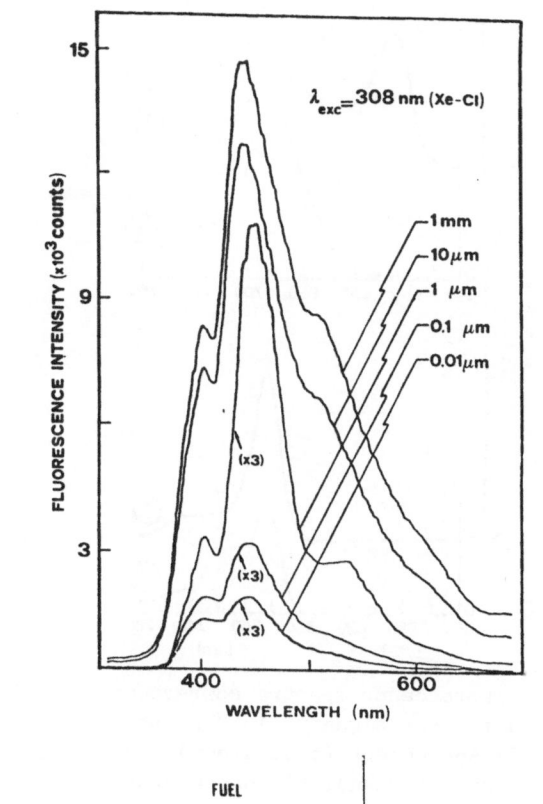

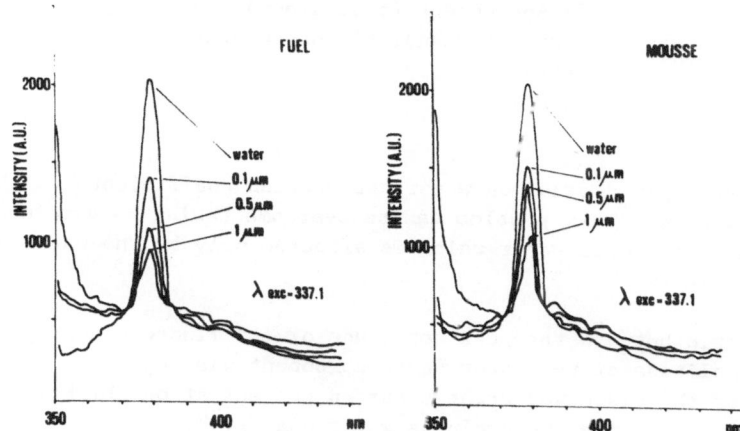

Fig. 12. Oil film over water. Influence of the film thickness on
the fluorescence spectrum (top) and on the water Raman
peak (bottom).

of remote sensing of pollutants which are solved or suspended in water or
floating over the water surface. Since many pollutants have a good fluorescence efficiency, the Laser Induced Fluorescence (LIF) technique, with
its water penetration performances, is presently the only one having a
potential for the detection of pollutants in the water column.

With the same pollutant concentration in the water column the LIF

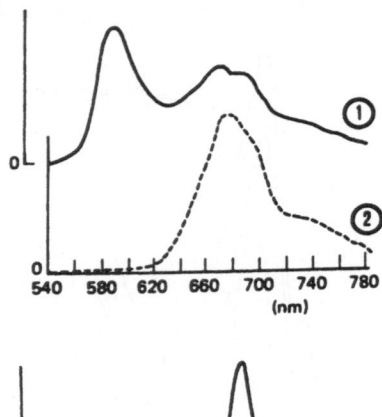

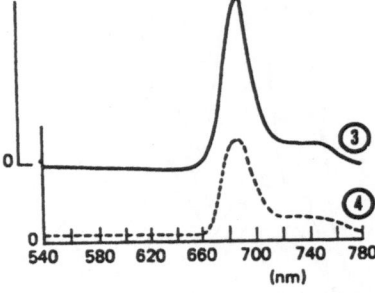

Fig. 13. Fluorescence spectra corresponding to
different algae[32]. 1) Rhodosorus (red),
2) Aphanotece (blue-green), 3) Eutrepia
marina (green), 4) Chaetoceros (golden-
brown).

intensity strongly depends on water attenuation coefficients which are in
principle unknown. This problem can be overcome taking as a reference the
Raman signal of liquid water which is affected only by the water attenuation coefficients.

The ratio between the peak amplitude of the fluorescence spectrum and
the peak amplitude of the water Raman component was suggested as a good
indicator of the dissolved organic carbon concentration which is an important parameter in order to evaluate water quality[30].

For what concerns the detection of specific pollutants, good results
were achieved in the monitoring of lignine sulfonate[31], a by-product of
wood industries, and of humic acids.

IV.4. Phytoplankton

Chlorophyll plays a fundamental role in the vegetation life, including phytoplankton, it is therefore obvious that each phenomenon related to
chlorophyll may be also related to vegetation and its health.

Phytoplankton shows a broad peak of chlorophyll fluorescence around
685 nm, which is used in laboratory in order to achieve a quick evaluation

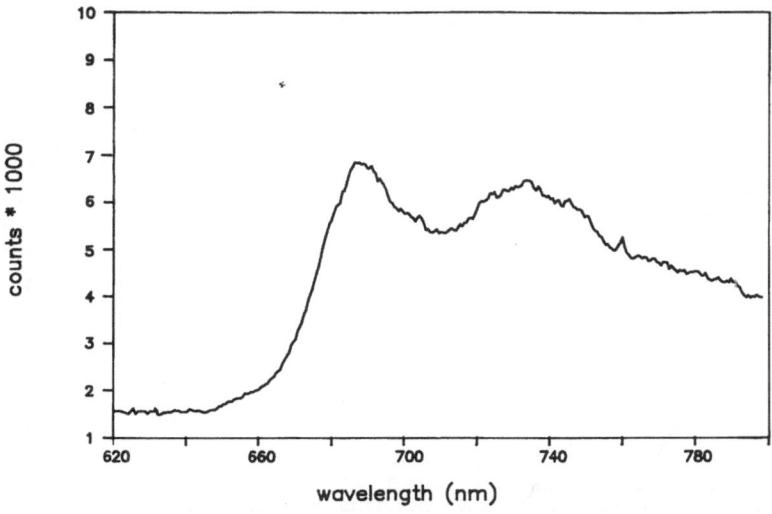

Fig. 14. Fluorescence spectrum of chlorophyll detected
with FLIDAR-2 on Douglas firs during the LIFT
87 experiment (Garderer, NL, June 1987).
Excitation wavelength 480 nm.

of the chlorophyll concentration in water samples. In field experiments
the phytoplankton is suspended in the water column and the water Raman
signal has to be taken as reference in order to evaluate the influence of
the water attenuation coefficients.

Additional fluorescence bands may appear because of particular fluorescent pigments which are characteristic of different phytoplankton
species[32] (Fig. 13). These bands mey be used as signatures for the
identification of the species if a spectral resolved receiver is used.

V. REMOTE SENSING OF VEGETATION

The remote sensing of LIF in vegetation is one of the most recent
Lidar techniques[11] and shows an interesting potential in vegetation
analysis. The fluorescence spectrum of chlorophyll _a_ in plants is given by
the convolution of two broad peaks having the maxima respectively at 685 nm
and 730 nm (Fig. 14). These peaks show a strong connection with the two
photoreacting systems which play a foundamental role in the photosynthetic
process. Therefore it is possible to derive the behavior of photosynthesis
from the fluorescence spectrum. In this way the changes in plant health
state can be monitored earlier than by other remote sensing techniques
which detect the changes of external characteristics, like the infrared
spectral reflectance.

Because of its direct connection to the photosynthetic process,
chlorophyll fluorescence depends on both the health state of the plant and

145

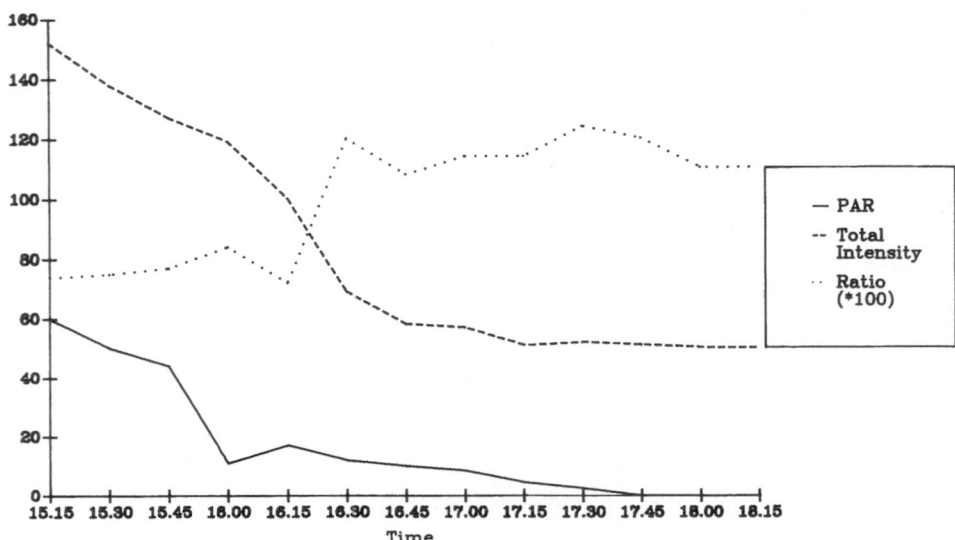

Fig. 15. Photosynthetic Active Radiation (PAR), total fluorescence,
and ratio between the fluorescence intensity at 730 nm and
685 nm as a function of time. Measurements done on Douglas
firs with the FLIDAR-2 during the LIFT 86 experiment
(Wageningen, NL, October 1986). Excitation wavelength
480 nm.

the solar radiation intensity[33]. A separate measurement of the photosynthetic active radiation (PAR) impinging on the plant is then suitable.

Both laboratory and field experiments have shown[34] the influence of
stresses on the ratio between the two peaks at 685 nm and 730 nm and
generally the correlation between this ratio and photosynthesis efficiency.

VI. CONCLUSIONS

Fluorescence lidars show a large potential in the remote sensing of
natural waters and vegetation. Beside the well established applications,
like oil remote sensing, new interesting applications have been recently
studied, particularly in vegetation monitoring. In this field the possibility of a direct analysis of the photosynthesis efficiency seems able to
detect stresses and deseases earlier than other remote sensing techniques.

REFERENCES

1. W. EK. Middleton and A. F. Silhaus, "Meteorological Instruments",
    University of Toronto Press, Toronto (1953)
2. F. J. Mc. Clung and R. W. Hellwarth, J. Appl. Phys., 33, 828-829
    (1962)
3. G. Fiocco and L. D. Smullin, Nature, 199, 1275-1276 (1963)

4. M. G. H. Ligda, Proc. Conf. Laser Technol., 1st, San Diego Cal., 63-72 (1963)

5. G. D. Hickman and J. E. Hogg, "Remote Sensing of the Environment", 1, 47-58 (1969)

6. R. M. Measures and M. Bristow, Can. Aeron. Space J., 17, 421-422 (1971)

7. J. F. Fantasia, T. M. Hard and H. C. Ingrao, "An Investigation of Oil Fluorescence as a Technique for Remote Sensing of Oil Spills", Report No. DOT-TSC-USCG-71-7, Transportation Systems Center, Dept. of Transportation, Cambridge, Mass. (1971)

8. W. Wiesemann, R. Beck, W. Englisch and K. Gurs, Appl. Phys., 15, 257-260 (1978)

9. F. E. Hoge, R. N. Swift and J. K. Yungel, Appl. Opt., 22, 2991-3000 (1983)

10. L. Pantani and I. Pippi, Optica Acta, 30, 1473-1481 (1983)

11. G. Cecchi, L. Pantani, I. Pippi, R. Magli and P. Mazzinghi, "Vegetation remote sensing: a new field for Lidar applications", ECOOSA '84, SPIE Vol. 492, 180-185, SPIE, Bellingham Wa. (1985)

12. E. D. Hinkley (Editor): "Laser monitoring of the atmosphere", Springer Verlag, Berlin (1976)

13. R. M. Measures, "Laser Remote Sensing", John Wiley and Sons, New York (1983)

14. G. W. Kattawar and G. N. Plass, Appl. Opt., 15, 3166-3178 (1976)

15. L. Pantani, Atti Fondazione Giorgio Ronchi, 36, 292-304 (1981)

16. R. M. Measures, W. R. Houston and M. Bristow, Can. Aeron. Space J., 19, 501-506 (1973)

17. R. A. O'Neil, L. Bujia-Bijunas and D. M. Rayner, Appl. Opt., 19, 863-870 (1980)

18. M. Bristow, D. Nielsen, D. Bundy and R. Furtek, Appl. Opt., 20, 2889-2906 (1981)

19. F. E. Hoge, R. N. Swift and E. B. Fredrick, Appl. Opt., 19, 871-883, (1980)

20. F. E. Hoge and R. N. Swift, Appl. Opt., 20, 1191-1202 (1981)

21. D. Diebel-Labgohr, R. Hengstermann, R. Reuter, G. Cecchi and L. Pantani: Measuring Oil at Sea by Means of an Airborne Laser Fluorosensor, in: "The Archimedes 1 Experiment", R. A. Gillot and F. Toselli Editors, Commission of the European Communities, EUR 10216 EN, ECSC-EEC-EAEC, 123-142, Brussels (1985)

22. F. Castagnoli, G. Cecchi, L. Pantani, I. Pippi, B. Radicati and P. Mazzinghi, Fluorescence Lidar for land and sea remote sensing, in: "Laser Radar Technology and Applications", J. M. Cruickshank and R. C. Harney Editors, SPIE Vol. 663, 212-216, Bellingham (1986)

23. G. Cecchi and P. Mazzinghi, "Un laser a scarica trasversa, in specie del tipo eccimero", Brevetto d'invenzione industriale n. 9552, 24/11/1987

24. J. M. Massin, Ed., "Remote Sensing for the Control of Marine Pollution", Plenum Press, New York (1984)

25. R. A. O'Neal, F. E. Hoge and M. P. F. Bristow, "The Current Status of Airborne Laser Fluorosensing", 15th International Symposium on Remote Sensing of the Environment, 379-389, Ann Arbor, Michigan (1981)

26. G. Cecchi, L. Pantani, P. Mazzinchi and A. Barbaro, Fluorescence Lidar Remote Sensing of the Environment: Laboratory Experiments for the Characterization of Oil Spills and Vegetation, in: "Optoelectronics in Engineering", W. Waidelich Ed., 652-655, Springer Verlag (1986)

27. R. M. Measures, H. R. Houston and D. G. Stephenson, Optical Engineer., 13, 454-494 (1975)

28. P. Burlamacchi, G. Cecchi, P. Mazzinghi and L. Pantani, Appl. Opt., 22, 48-53 (1983)

29. G. Cecchi, P. Mazzinghi, L. Pantani and C. Susini, Lidar Investigation of Oil Films on Natural Waters, in: "Optoelectronics in Engineering", W. Waidelich Ed., 517-522, Springer Verlag, Berlin (1984)

30. M. Bristow, D. Nielsen, "Remote Monitoring of Organic Carbon in Surface Waters", Report No. EPA-80/4-81-001, U.S. Environmental Protection Agency, Las Vegas, Nevada (1981)

31. M. Bristow: Remote Sensing of Environment, 7, 105-127 (1978)

32. P. B. Mumola, O. Jr. Jarrett and C. A. Jr. Brown, "Multiwavelength Lidar for Remote Sensing of Chlorophyll a in Algae and Phytoplankton", NASA Conference on the Use of Lasers for Hydrographic Studies, NASA SP-375, 137-145 (1973)

33. G. Cecchi, C. Fagotti and P. Paoli, Misure di Fluorescenza della clorofilla in vivo, Quaderni Metodologici n. 4, CNR-IPRA, Roma (1985)

34. F. Castagnoli, G. Cecchi, L. Pantani, B. Radicati, M. Mazzinghi, A. Barbaro and M. Romoli, "A New Lidar Sytem for Applications Over Land and Sea", Proc. of the 4th International Colloquium on Spectral Signatures of Objects in Remote Sensing, ESA SP-287, 239-243, Paris (1988)

HYDROGRAPHIC APPLICATIONS OF AIRBORNE LASER SPECTROSCOPY

R. Reuter

University of Oldenburg
Physics Department
Oldenburg, Federal Republic of Germany

I. INTRODUCTION

In the past decade remote sensing methods have become an important
tool in oceanographic research. Utilization of optical radiometers onboard
satellites for measurements of the colour of the sea has led to remarkable
results over the open ocean, yielding global maps of chlorophyll content
in the surface layer[1,2]. Problems arise in the interpretation of ocean
colour data obtained over coastal waters due to the presence of other
materials such as suspended mineral particles and dissolved organics
(Gelbstoff) which also contribute to the radiometer signal.

Another remote sensing method makes use of lidar systems installed in
aircraft[3,4]. The airborne hydrographic lidar is an active sensor consisting
of a laser emitting short pulses at near UV or visible wavelengths which
are deflected towards the water surface, and of a gated signal receiver for
the detection of laser-induced radiation. Compared to passive radiometry,
two main differences are obvious: (1) the method is not dependent on sun-
light-operation at night yields an even better signal-to-noise level since
daylight background does not interfere with the laser-induced signals; (ii)
owing to the use of monochromatic laser light and registration of scattering and fluorescence it is possible to gather specific information on
optical parameters or fluorescent substances in the upper water layers.

A nearly synoptic survey of extended coastal areas can be achieved
using these instruments. For example, mapping of the hydrograpic situation
in the German Bight requires a flight time of 2 hours which is essentially
shorter than the tidal period. A close distance of typically 100-300 m
between the instrument and the water surface reduces atmospheric effects
significantly. Laser spectroscopy enables specific measurements of the
turbidity of water[5], and of the concentrations of Gelbstoff[5] and chlorophyll $\underline{a}$[3,4]. The elastic backscatter signal contains additional information on the composition of the hydrosol; however, only a few attempts have

been made at its registration because of the interference with laser light reflected from the sea surface.

In addition to the investigation of these parameters of interest in oceanographic research, lidar allows monitoring of marine oil pollution[6-9]. Due to the specific optical properties of mineral oils, a classification of the oil type[10,11] and a measurement of oil film thickness in the micrometre range[12,13] can be achieved. Detection of submerged oil volumes should be possible in principle, and some results of airborne experiments can be interpreted with signals from oil floating below the sea surface, but a demonstration under controlled conditions has not been achieved so far.

A main characteristic of lidar as an optical radar is its capability of deriving range-resolved data from time-resolved measurements of the signal return. Range resolution is a function of the laser pulse width and the bandwidth of the detection system. Depth profiles of turbidity were obtained for the first time in 1984 in the northern Adriatic down to a water depth of 17 m which corresponds to six attenuation lengths[14]. However, the depth-resolving mode can be successfully applied only under dark conditions, since daylight background contaminates the weak signal contributions detected from deeper water layers. The results presented here are derived from depth-resolved data which have been numerically integrated. This is equivalent to the data of a depth-integrating laser fluorosensor.

II. THE OCEANOGRAPHIC LIDAR SYSTEM

In this report, some results obtained with the Oceanographic Lidar System (OLS) of the University of Oldenburg are reviewed. This instrument was developed in 1983-84 with support from the Federal Minister for Research and Technology, Bonn, and operated in DO 28 and DO 228 research aircraft of DLR Oberpfaffenhofen.

The instrument has been specifically designed for oceanographic research. Maps of water column parameters in coastal zones can be almost synoptically obtained since measuring times are short as compared to the characteristic time scales of hydrographic changes, as, for example, the tidal period. A second area of application that has been studied in detail is the fluorometric analysis of oil spills in turbid coastal waters where strong fluorescence contributions from natural seawater compounds are also present.

A XeCl excimer laser emitting at 308 nm serves as the main light source. Front and rear laser output are utilized as lidar beam or as pumping beam for a dye laser tuned to an emission at 450 nm. Signal receiver is a Schmidt-Cassegrain telescope. Its optical axis is almost coaxial with the laser beams, and the 5 mrad field of view corresponds to the excimer laser beam divergence. Dichroic beamsplitters deflect selected spectral ranges of the telescope output to optical filters and gated photomultipliers. Three selectable signals are sequentially combined on one signal line and fed to a one-channel fast transient recorder.

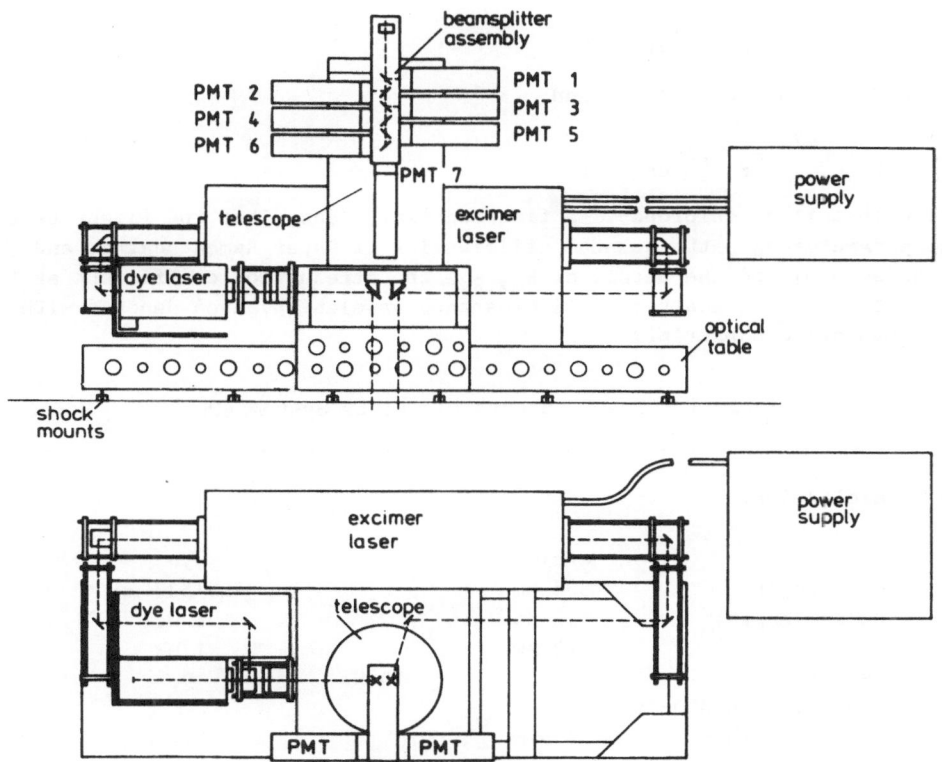

Fig. 1. Optical part of the Oceanographic Lidar System. Position
of the telescope is above a bottom hatch of the aircraft
for free field of view to the water surface.

Lasers and the detector systems are mounted on an optical table to
obtain a rigid alignment of the optical setup (Figure 1). System operation
is done in-flight by one operator. A microcomputer controls laser selection, triggering and power, selection of different detection wavelengths,
quick look data output and auxiliary functions such as photocamera activa-
tion and recording of global radiation data. Other technical data of the
instrument are listed in Table 1.

III. HYDROGRAPHIC MEASUREMENTS

Monochromatic irradiation of natural seawater yields spectral structures (Figure 2), which are related to (i) Rayleigh and Mie scattering of
molecules and particles, (ii) Raman scattering of water with a wave-number
shift of 3400 cm^{-1}, (iii) fluorescence of Gelbstoff (dissolved organic
matter), and (iv) fluorescence of chlorophyll a. Assuming an optically
deep and homogeneous water column, and starting from the hydro-graphic
lidar equation, the interpretation of time-integrated signals returns[4]
obtained from the inelastic spectral structures (ii)-(iv) yields a
detected power

$$P_R \sim H^{-2} \; \sigma_R \; 1/(k_{ex} + k_R)$$

for water Raman scattering, and

$$P_F \sim H^{-2} \; \sigma_F \; 1/(k_{ex} + k_F)$$

for Gelbstoff or chlorophyll a fluorescence. H describes the flight height, the parameter $\sigma_{R,F}$ the quantum efficiencies of water Raman scatter and fluorescence, and the parameter $k_{ex,R,F}$ the attenuation coefficient at the excitation, Raman scatter of fluorescence wavelength, each denoted with the respective subscript.

Table 1. The Oceanographic Lidar System (OLS)

excitation:

lasers:	excimer	dye
wavelength:	308 nm	450/533 nm
pulse length:	12 ns	6 ns
peak power:	10 MW	1 MW
rep. rate:	10 Hz	
footprint at 800 ft		
flight height:	2.5 m	0.7 m

detection:

telescope:	f/10 Schmidt-Cassegrain, diam. 0.4 m
wavel. selection:	dichroic splitters, interference filters and blocking filters
wavelengths (nm):	344 water Raman - λ_{ex} = 308 nm and Gelbstoff background
	366 Gelbstoff fluorescence
	380 Gelbstoff fluorescence
	500 Gelbstoff fluorescence
	533 water Raman - λ_{ex} = 450 nm
	650 Gelbstoff fluorescence
	685 Chlorophyll a fluorescence
detectors:	photomultipliers EMI 9812/9818, gated
digitizer:	Biomation 6500, 500 MHz, 6 bit (in the depth resolving lidar mode a fast logarithmic amplifier with 10 ns/decade fall time to enhance the dynamic range).
system computer:	LSI 11/23 with floppy disk, hard disk and magnetic tape
total weight:	500 kg

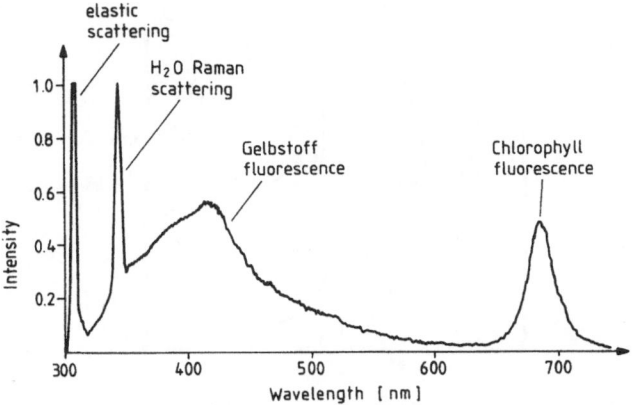

Fig. 2. Emission spectrum of a water sample taken from
the German Bight. Excitation wavelength is
308 nm. Gelbstoff represents dissolved organic
matter in seawater, being of natural origin and
brought to sea by river run-off. Chlorophyll a
fluorescence is used to determine phytoplankton.

σ_R can be set to be approximately constant, and the Raman signal intensity reflects the inverse sum of attenuation coefficients at the laser and the Raman wavelength. Using the two lasers installed in OLS with 308 and 450 nm emission, water Raman signals are observed at 344 and 533 nm, and the following data are derived:

UV   attenuation coefficient = $k(308) + k(344) \sim 1/P_R(344)$

VIS attenuation coefficient = $k(450) + k(533) \sim 1/P_R(533)$.

Gelbstoff or chlorophyll fluorescence signal power P_F read:

$$P_F \sim \sigma_F/(k_{ex} + k_F),$$

which after normalizing to the water Raman scatter signal yield:

$$P_F/P_R = \sigma_F/\sigma_R \, (k_{ef} + k_R)/(k_{ef} + k_F).$$

In this equation, σ_F is the only variable parameter if the ratio $(k_{ex} + k_R)/(k_{ex} + k_F)$ can be set to be a constant. This requires a spectrally close selection of excitation and detection wavelengths. Fluorescent matter concentrations derived with OLS then read:

Gelbstoff   = $F(366)/R(344)$

chlorophyll = $F(685)/R(533)$

obtained with 308 and 450 nm excitation, respectively. Although being given in dimensionless units, these parameters represent absolute

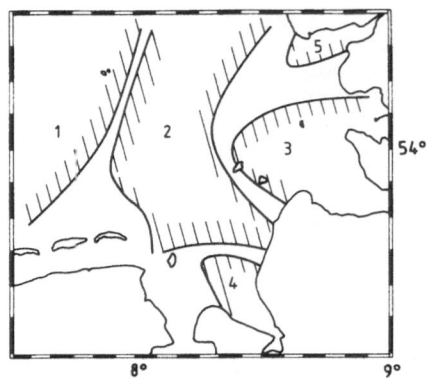

Fig. 3. Characteristic water types in the German
Bight derived from airborne Gelbstoff
fluorescence and shipboard salinity data
on October 3-5, 1985. Type 1 represents
open sea water, type 2 is produced by
mixing of open sea and river water, and
types 3, 4 and 5 correspond to run-off of
rivers Elbe, Weser and Eider, respectively.

quantities in terms of efficiencies of fluorescent compounds normalized to
the constant water Raman efficiency.

As an example of the oceanographic campaigns performed with OLS,
some results obtained in the experiment "Fronten II" in October 1985 in
the German Bight are presented. The experiment aimed at an investigation
of characteristic water masses and fronts in this area and was jointly
performed by oceanographers from the Alfred Wegener Institute for Polar
and Marine Research, Bremerhaven, and from institutes of the University of
Hamburg, and by the remote sensing group of the University of Oldenburg.

It could be shown in this experiment that the distribution of
Gelbstoff obtained with OLS provides an efficient method to describe river
plume fronts in the German Bight caused by the freshwater input of rivers
Elbe and Weser. Two fronts of this type were identified, their position
and the location of water masses separated by these fronts are shown in
Figure 3. This map has been derived from data obtained from four flights
made over a period of 3 days, and with the support of shipboard ground
truth. The lower limit of sensitivity of the airborne Gelbstoff measurement is almost the same as that obtained with laboratory instrumentation.
Airborne Gelbstoff data were compared with salinity measured in the
surface layer at identical ship and aircraft positions, whereby the
difference of the sampling time of data chosen for this purpose did not
exceed 1 hour. The covariance of these parameters is high, as is typically
found in the German Bight, with a correlation coefficient of -0.96.

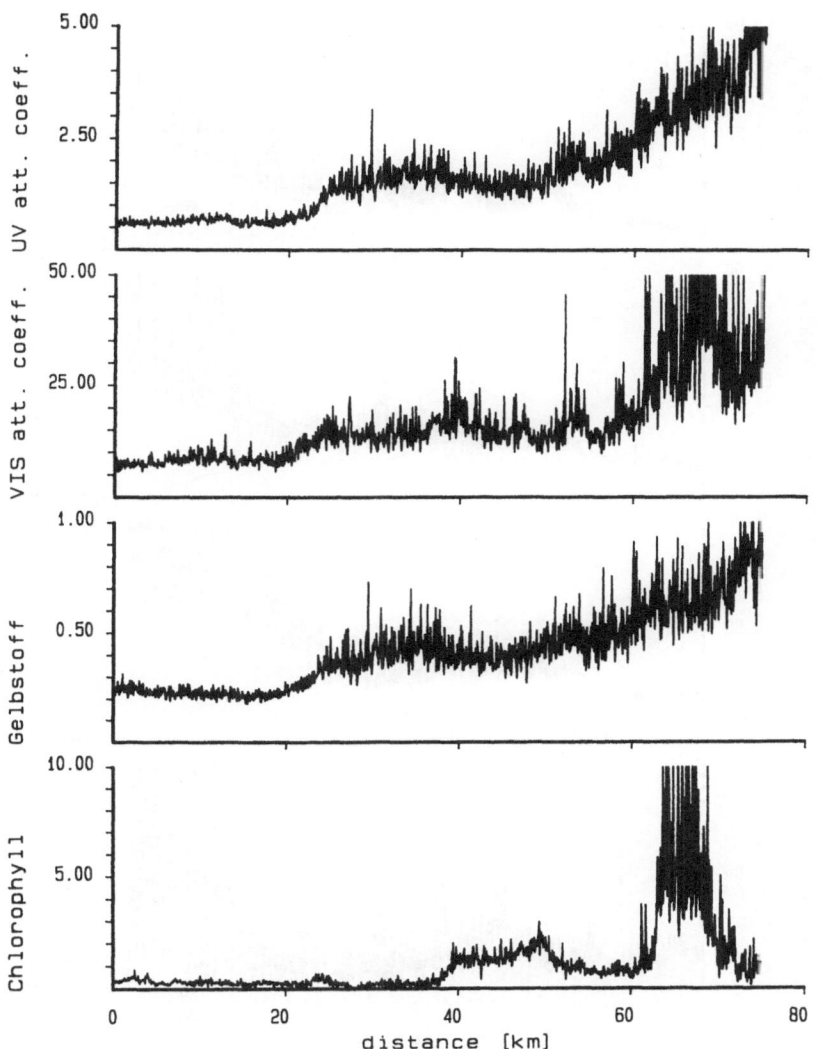

Fig. 4. West-east section on 54°15'N from 7°40'E to 8°50'E, October 4, 1985 (flight number 083, time 12:47-13:08). The front separating water types 1 and 2 is located around km 22, being about 7 km wide. Water type 5, river Eider run-off, is marked by a drastic increase in Gelbstoff and turbidity, and a strong plankton bloom is identified in this area from the high chlorophyll data.

To further describe the hydrographic situation derived with OLS, profiles of the UV and VIS attenuation coefficients, and Gelbstoff and chlorophyll obtained on October 4, 1984 on longitude 54°00'N and 54°15'N are shown in Figures 4 and 5. The formation of water types 1, 2, 3 and 5 separated by frontal transition areas can be clearly seen in the Gelbstoff data. According to shipboard ground truth, chlorophyll concentration is low, being of the order of a few micrograms per litre, and this parameter obviously did not correlate with the other parameters measured.

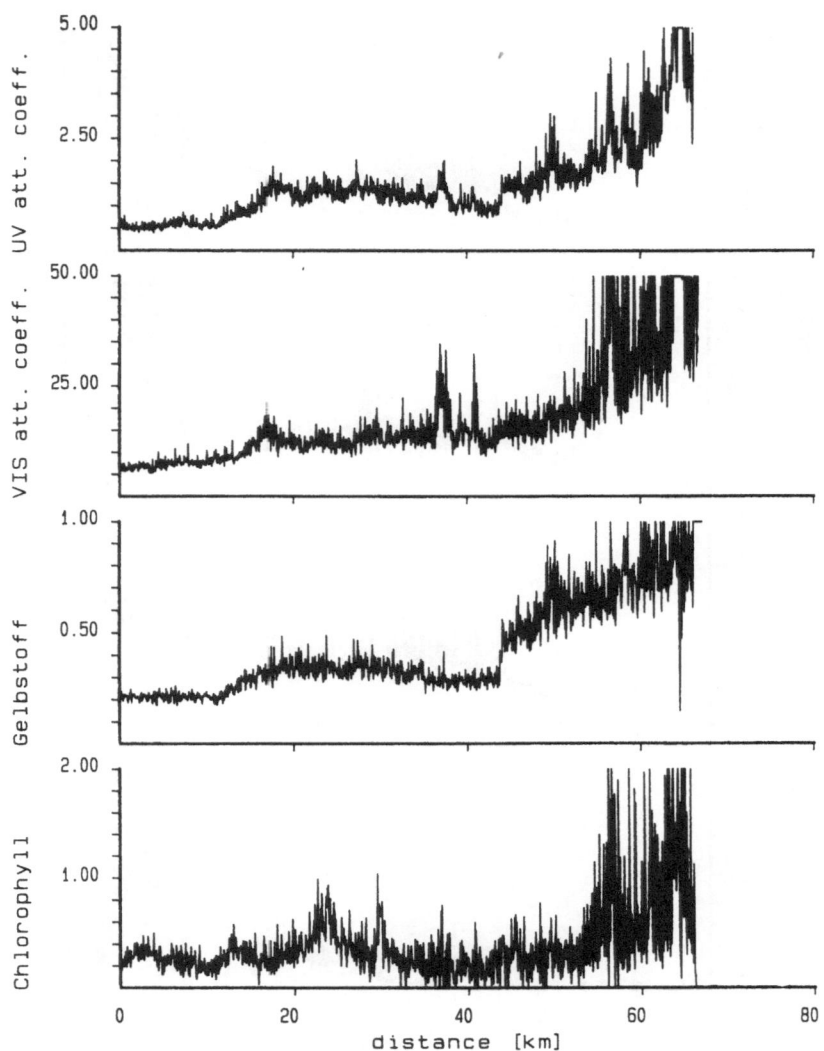

Fig. 5. West-east section on 54°00'N from 7°40'E to 8°45'E, October
4, 1985 (flight number 081, time 11:58-12:22). The fronts
separating water types 1 and 2, and water types 2 and 3, are
located around km 13 and km 43, respectively.

IV. INVESTIGATION OF OIL SPILLS

It has been the goal of airborne campaigns performed with OLS to
investigate the capabilities of deriving information on oil spills in
cases typically found following controlled discharges by ship traffic or
off-shore platforms. Then, the amount of oil is mostly low when compared
to accidental spills, giving rise to a film thickness of the order of only
a few micrometers.

This situation was simulated in the MARPOL exercise[9] northwest of
Rotterdam on May 7, 1986, in cooperation with Rijkswaterstaat, The

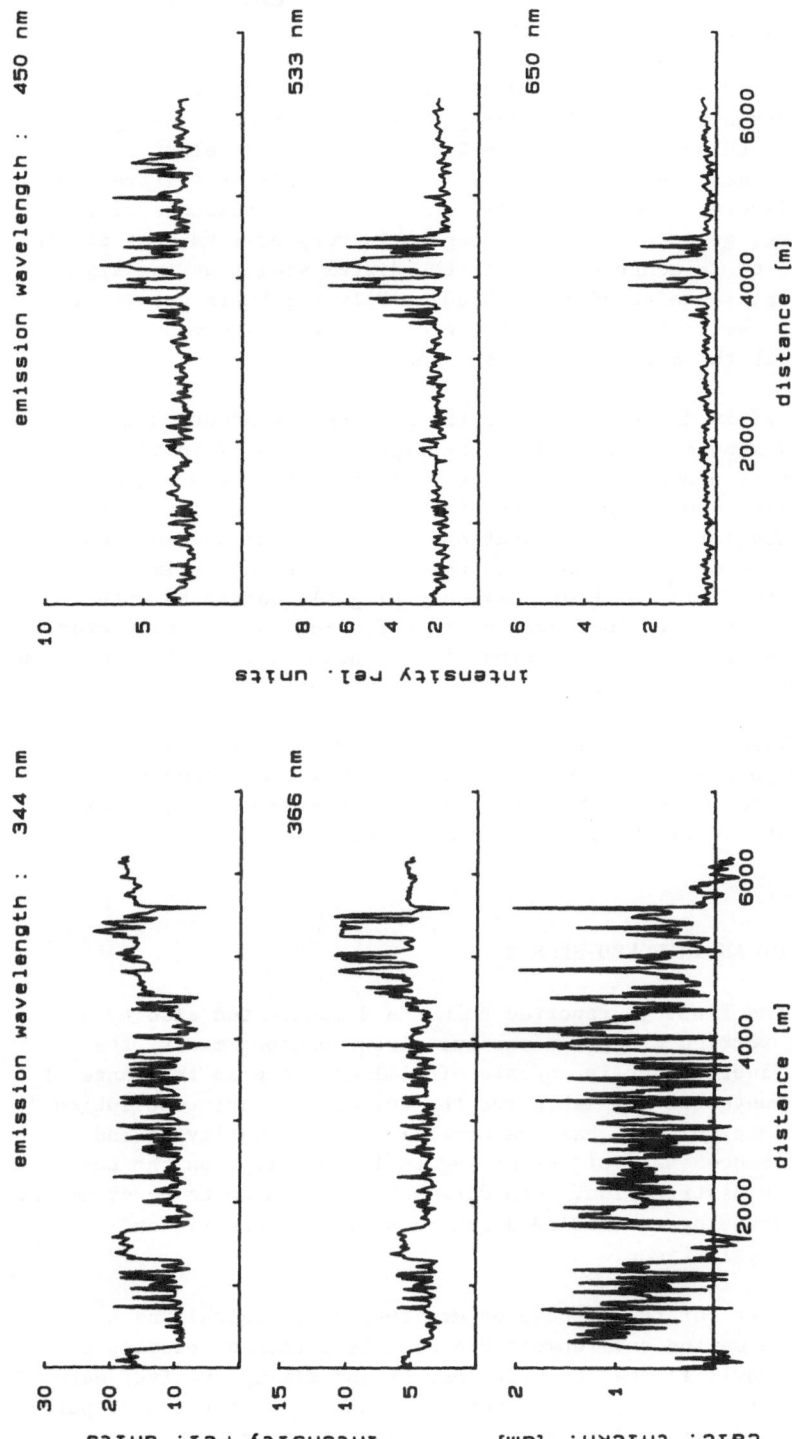

Fig. 6. Profile obtained at various emission wavelengths from a crude oil and diesel spill, both discharged in a quantity of 60 liters/nautical mile. Crude oil begins at position 500 m. Crude oil ends, and diesel fuel begins at position 4600 m. Ship position is at 5700 m. The profile of film thickness calculated from these data is displayed on the left, lowest trace. (Date: May 7, 1986; time: 7:55; excitation wavelength: 308 nm).

157

Netherlands. A moving vessel discharged 60 liters of oil per nautical mile, mixed with 100 m^3 of seawater before spilling. Two types of oil were utilized: a mixture of gasoil, fuel, lubrication and crude oil (denoted here as crude oil) and a diesel fuel.

Results of an OLS overflight 30 minutes after beginning crude oil discharge, and 3 minutes after beginning diesel discharge, are shown in Figure 6. Outside the oil spill, water Raman scatter and Gelbstoff fluorescence are measured. Over the spill these signals are altered by absorption and fluorescence of the oil. The strong fluorescence of the crude oil at blue, green and red wavelengths is very obvious near distance 4000 m in Figure 6. In contrast to this the diesel yields strong signals at UV wavelengths, shown at distance 5000 m. Film thickness derived from the depression of water Raman scatter yields data that are considered to be consistent with the amount of oil spilled.

Data obtained in an overflight of the same spills about 30 minutes later (Figure 7) show a depression of the signal intensity in all detection channels, more pronounced in the UV than at higher wavelengths. The characteristic fluorescence signature of the oils is lost. This effect is most probably associated with evaporation of fluorescent oil compounds due to windspeeds of about 5 bft. However, the bulk of the oil films is still present: calculated film thickness has only slightly decreased when compared to the data shown in Figure 6. This finding is also supported by visual inspection: a marked alteration of the spills could not be made out in the period of the overflights.

It is concluded that the fluorescent compounds of mineral oil are affected to a higher extent than other surface-active components. This limits fluorometric fingerprinting of small discharges of mineral oil if identification of the oil type is of primary interest.

V. DEVELOPMENT OF AN IMPROVED SENSOR

Following the findings reported here and demonstrated also by a number of other working groups in various airborne experiments, the airborne laser fluorosensor is capable of filling a gap in the range of airborne instrumentation available for the purpose of marine pollution monitoring. In particular it has the capability for identifying and classifying substances, and of estimating their quantity, on the basis of film thickness distribution. Such data have been shown to be of utmost relevance for surveying controlled discharges by ship traffic and off-shore oil installations.

Data important for the purpose of monitoring biological and physical parameters of the marine environment can also be obtained: biomass production from chlorophyll fluorescence, transport and mixing of river water in the coastal zone from Gelbstoff fluorescence, and concentration of particulate and dissolved material from water Raman scattering yielding turbidity data. In addition to this, the potential exists for the identifica-

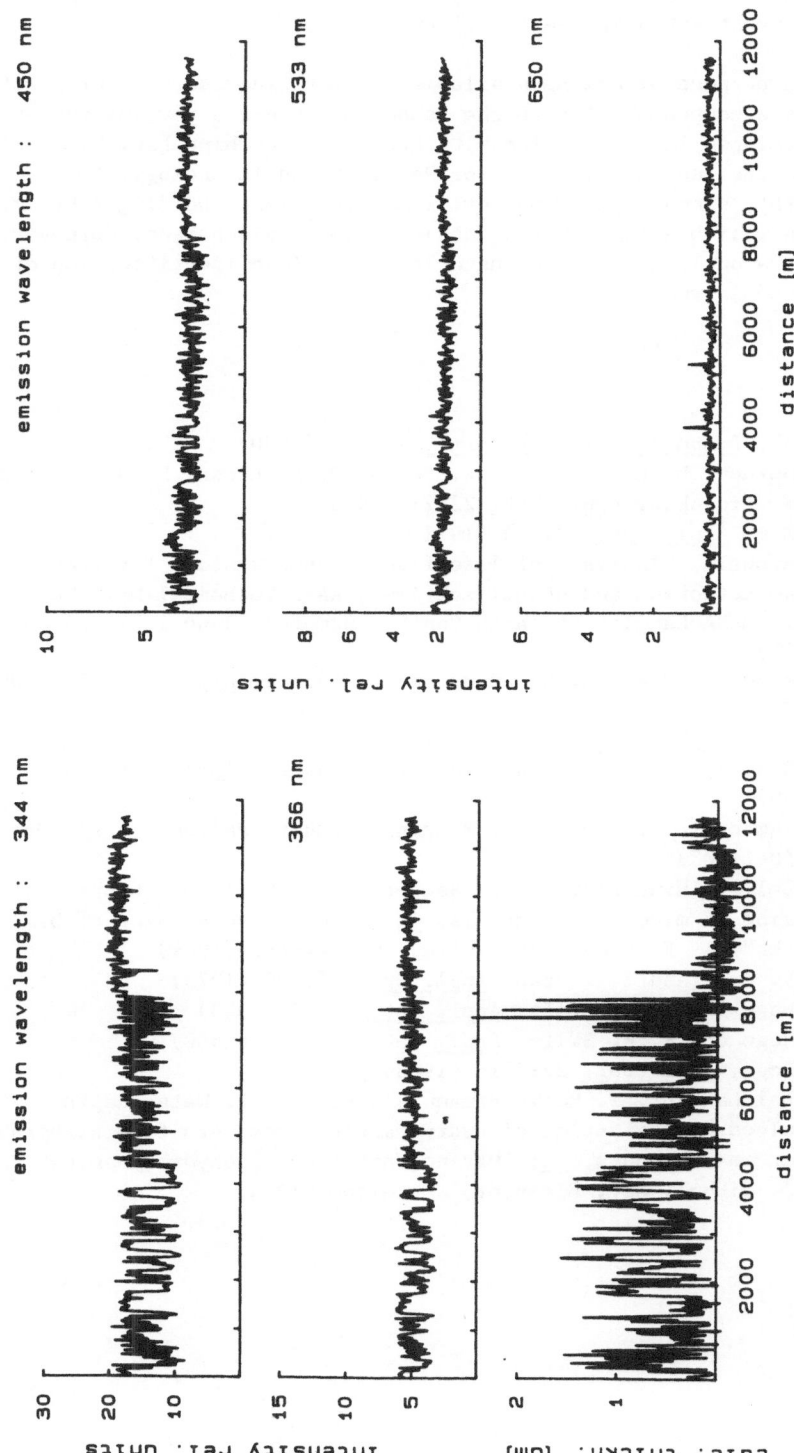

Fig. 7. Same situation as shown in Fig. 6, 30 min. later. Crude oil begins at position 200 m. Crude oil ends, and diesel fuel begins at position 4500 m. Diesel fuel ends at position 8000 m. (Date: May 7, 1986; time 8:25; excitation wavelength: 308 nm).

tion of chemical pollutants when discharged in larger quantities, e.g., following ship accidents or in highly polluted estuaries. This topic will be further investigated in the near future.

For the purpose of making available a laser fluorosensor for routine operation as a component of airborne sensor packages, a new instrument has now been developed in cooperation with Krupp MaK Maschinenbau, Kiel, with support from the Federal Minister for Research and Technology, Bonn. Equipped with a scanner allowing aerial mapping along the flight track, it will provide reliable data on the total volume of discharges. Moreover, in-flight data analysis will be installed to achieve identification of the substance type in real time.

REFERENCES

1. A. Morel, Boundary Layer Meteorol. 18:177 (1980)
2. H. R. Gordon, D. K. Clark, J. W. Brown, O. B. Brown, R. H. Evans and W. W. Broenkow, Appl. Opt. 22:21 (1983)
3. H. H. Kim, Appl. Opt. 12:454 (1973)
4. E. V. Browell, "Analysis of laserfluorosensor systems for remote algae detection and quantification", NASA Technical Note TN D-8447, NASA Langley Research Center, Hampton, June 1977, 39 pp. (1977)
5. M. Bristow, D. Nielsen, D. Bundy and R. Furtek, Appl. Opt. 20:2889 (1981)
6. H. Visser, Appl. Opt. 18:1746 (1979)
7. R. A. O'Neil, L. Buja-Bijunas and D. M. Rayner, Appl. Opt. 19:863 (1980)
8. P. Burlamacchi, G. Cecchi, P. Mazzinghi and L. Pantani, Appl. Opt. 22:48 (1983)
9. D. Diebel, T. Hengstermann, R. Reuter and R. Willkomm, Laser Fluoro-sensing of mineral oil spills, in: "The Remote Sensing of Oil Slicks", A. E. Lodge, ed., Wiley, Chichester (1989)
10. R. T. V. Kung and I. Itzkan, Appl. Opt. 15:409 (1975)
11. F. E. Hoge and R. N. Swift, Appl. Opt. 22:37 (1983)
12. F. E. Hoge and R. N. Swift, Appl. Opt. 19:3269 (1980)
13. F. E. Hoge, Appl. Opt. 22:3316 (1983)
14. D. Diebel-Langohr, T. Hengstermann and R. Reuter, Water depth resolved determination of hydrographic parameters from airborne lidar measurements, in: "Marine Interfaces Ecohydrodynamics", J. C. J. Nihoul, ed., Elsevier, Amsterdam (1986)

HEAVY METAL ANALYSIS

QUANTITATIVE DETECTION OF STRONTIUM-90 AND STRONTIUM-89

IN ENVIRONMENTAL SAMPLES BY LASER MASS SPECTROMETRY

K. Wendt, G. Haub, S. Köhler, H.-J. Kluge, L. Monz,
E.-W. Otten, G. Passler, P. Senne, J. Stenner, K. Zimmer*
G. Herrmann, N. Trauttmann and K. Walter**

  *Institut für Physik
 **Institut für Kernchemie
   Universität Mainz
   Mainz, Federal Republic of Germany

I. INTRODUCTION

Parallel to the strongly growing public concern about environmental
problems, new ideas for trace detection and analysis of toxic and radioactive material are being developed. One of these new and outstanding
experimental techniques is the application of analytical laser spectroscopy. Most interesting in this context is the method of resonance ionization spectroscopy (RIS), as proposed[1] already in 1972 combining very high
sensitivity in the detection of the element or isotope under investigation
with high selectivity in the suppression of contaminants[2-4].

A challenge for trace analysis is the fast and sensitive detection in
environmental samples of the neutron rich strontium isotopes with mass
number A = 89 and A = 90. Both radioactive isotopes do not occur naturally
but are produced man-made as fission products in nuclear weapon tests or
in case of a reactor accident. The long-lived isotope ^{90}Sr ($T_{1/2}$ = 28.5 y)
is particularly dangerous as it is accumulated in human bones, being
chemically equivalent to calcium. Accordingly, the incorporation of a
trace contamination of as little as 37 kBq (= 1 µCurie), corresponding to
10^{13} atoms, is already supposed to be a dangerous exposure. This number
has to be compared with the typical fall-out of 10^8-10^{12} atoms per square
meter which was measured in Western Europe[5-7] shortly after the Chernobyl
reactor accident in 1986.

The radiochemical detection of both isotopes ^{89}Sr and ^{90}Sr is
complicated due to the fact that ^{90}Sr is a pure β-emitter and thus cannot
easily be distinguished from the β-radiation of the short-lived neighbouring isotope ^{89}Sr ($T_{1/2}$ = 50.5 d). Therefore the radiochemical trace

Optoelectronics for Environmental Science, Edited by S. Martellucci and
A.N. Chester, Plenum Press, New York, 1990

detection of ^{90}Sr has to be carried out via a time-consuming separation of its radioactive daughter product ^{90}Y ($T_{1/2}$ = 54.1 h). Including sample preparation, chemical separation, and subsequent counting the full procedure requires about 14 days to achieve a reasonable sensitivity limit, corresponding to the detection of about 10^8 atoms per sample. As a consequence of that a radioactivity-independent, fast and quantitative detection method for trace amounts of ^{89}Sr and ^{90}Sr is desirable. An approach by optical techniques is hampered by the relatively high natural abundance of stable strontium isotopes, mainly ^{88}Sr. A typical sample, as extracted from 1000 m^3 air, contains up to 1 mg stable strontium, corresponding to about 10^{19} atoms, but only about 10^8-10^{10} atoms of ^{89}Sr and ^{90}Sr.

Thus the requirements that must be met by a radioactivity independent and competitive detection method, are as follows:

(i) the selectivity in the suppression of stable strontium isotopes must exceed a value of about 10^{11};

(ii) the detection efficiency in the trace analysis of only about 10^8 atoms must exceed a value of about 10^{-5} for reasonable statistics;

(iii) the total measuring time including sample preparation should not exceed one day.

Different groups have proposed a number of methods for ultrasensitive isotope analysis, especially suited for the detection of ^{89,90}Sr in the environment. Most of them are based on analytical laser spectroscopic techniques[8,9]. Furthermore, the use of accelerator mass spectrometry has been discussed.

In this paper a new technique for the detection of trace amounts of the radioactive isotopes ^{89}Sr and ^{90}Sr in environmental samples is presented. After chemical extraction, subsequent surface ionization, acceleration to 60 keV and mass separation in a magnetic sector field, the Sr-ions are charge exchanged into the metastable 5s5p ^{3}P$_2$-level. Resonance ionization with cw dye lasers in collinear geometry is applied for excitation to a Rydberg level, where the Rydberg atoms are field-ionized and detected. In this way an efficiency of better than 10^{-5} and a selectivity of better than 10^{11} in the suppression of the dominant stable strontium isotopes, mainly the isotope ^{88}Sr, is envisaged. Measurements are planned on environmental samples with concentrations of a minimum of 10^8 atoms of ^{89}Sr and ^{90}Sr per sample at a background of up to 10^{19} atoms of stable strontium in a measuring time of a few hours.

In this following we present a procedure to meet the requirements listed above for sensitive detection of ^{89,90}Sr in environmental samples, by combining a mass separator with highly efficient and highly selective on-line collinear fast-beam laser spectroscopy. Collinear laser spectroscopy was proposed by Kaufman[10]. It is presently applied in various experimental schemes[11,12] mainly for the study of the nuclear structure of short-lived isotopes. The following chapters will first focus on the general theoretical basis of a technique that could meet the requirements for both selectivity and efficiency simultaneously. Afterwards, the resulting experimental apparatus, presently under construction in our institute, will be described.

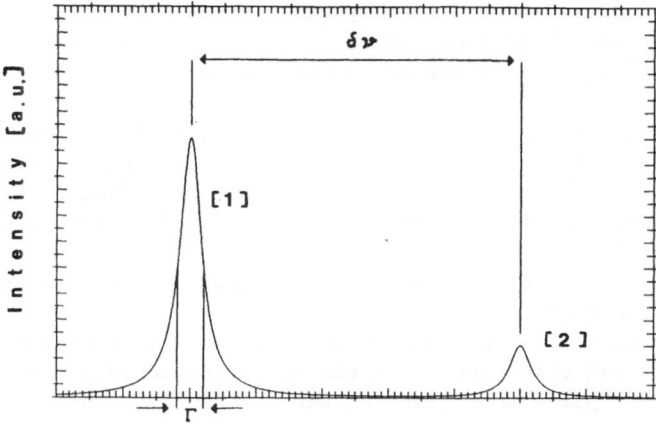

Fig. 1. Selectivity in laser spectroscopic experiments limited by the contribution of the
long tail of Lorentzian curve (1) to the
peak of Lorentzian curve (2).

II. SELECTIVITY IN THE LASER SPECTROSCOPIC EXPERIMENT

The selectivity of any experiment can be defined as the suppression
of an unwanted contaminant with respect to the quantity under study. In a
laser spectroscopic experiment involving a highly abundant isotope (1) and
a trace isotope (2) this ratio is usually given by the height of two
resonance peaks as shown in Fig. 1. Here the laser is tuned to the small
resonance ν_2, and the contamination is represented by the long tail of the
strong resonance at ν_1, which still might lead to a sizeable contribution
at the peak position of resonance ν_2.

The general shape of the resonance curve will usually be given by a
convolution of a Gaussian line profile, arising from inhomogeneous Doppler
broadening, and a Lorentzian line profile, arising from the homogeneous
natural line width. Nevertheless it can easily be shown that, for line
separations larger than ten times the line width, only the Lorentzian
shape contributes significantly and the Gaussian contribution is abso-
lutely negligible. In this case the selectivity is given by the simple
formula

$$S = \frac{H_2(\nu_2)}{H_1(\nu_1)} = \left(\frac{2\,\delta\nu}{\Gamma}\right)^2 \tag{1}$$

where $H_\nu(\nu_2)$ is the height of resonance (n) at the peak position of
resonance (2). Thus, in order to achieve a high value of S, the line
separation $\delta\nu$ must obsviously be much larger than the line width Γ. This
requirement applies even under ideal experimental conditions limited by
the natural line width due to the life time of the levels and the time-of-

flight-broadening, which is given by the maximum excitation time for one atom in passing the laser beam. It is therefore obvious that a maximum value for the frequency separation of the two resonances must be realized.

For typical values of the isotope shift and radiative line width even the application of narrow-band laser excitation results in selectivities of about only $S \cong 10^4 - 10^6$. For the realization of higher selectivities three general experimental principles are discussed in Ref. 2 (further references therein):

(i)   the use of repeated resonance interactions by multiple excitation and reradiation from each atom;

(ii)   the case of multistep resonance excitation and resonance ionization, where the selectivities S_1 in the different optical transitions used can be multiplied to obtain the total selectivity of the overall process

$$S = \pi \; S_1 ; \qquad\qquad\qquad\qquad\qquad\qquad (2)$$

(iii) the introduction of an "artificial" mass shift between different isotopes by utilizing a beam of accelerated atoms and taking advantage of the difference in the masses via the Doppler shift.

In addition to this, the application of these techniques to trace analysis is often realized with a combination of the different approaches mentioned above. A highly sophisticated example for the study of ^{90}Sr concentration is based on a combination of the first and second experimental principle. It is described in a proposal by Snyder and coworkers[9], who claim to reach selectivities exceeding 10^{15}. The group working with V. S. Letokhov has developed a number of other experimental techniques, based on the combination of different schemes as well as further mass separation steps. These include multiphoton ionization in a hot cavity with subsequent acceleration of the photo ions for mass separation, the so called "laser-ion-source"[13] (see also the contribution of F. Ames in this book), and resonance ionization in a fast atomic beam. The latter method has already been tested on rare isotopes of aluminum in 1982 by Kudriavtsev and Letokhov[14], and successfully applied[15] for the detection of trace amounts of ^{3}He (see also the contribution of V. S. Letohov to this book). This method overcomes the problem that most transitions between high-lying states do not show significant isotope shifts and thus yield no additional selectivity. In the case of the strontium isotopes in particular, the two contributions to the isotope shift between ^{88}Sr and ^{88,90}Sr, the field and mass shift, have opposite signs and partially cancel other other. Thus a similar approach, based on the introduction of the "artificial" mass shift seems to be feasible to achieve extremely high selectivity in the case of ^{89,90}Sr separation from ^{88}Sr.

III. DOPPLER WIDTH AND DOPPLER SHIFT IN COLLINEAR LASER SPECTROSCOPY

The illumination of a beam of fast atoms in the beam direction by a collinear laser beam leads to a number of special experimental features, all related to the Doppler effect. During electrostatic acceleration of an ion beam the energy spread δE remains constant due to conservation of phase space. This fact can be written as

$$\delta E_{k\ in} = \delta \left(\frac{1}{2}\ m\ v_z^2\right) = m\ v_z\ \delta\ v_z = \frac{m\ c^2}{v_0^2}\ \Delta\nu_D\ \delta\nu_D \tag{3}$$

for any resonance frequency ν_0. The formula shows that the introduction of a strong Doppler shift $\Delta\nu_D$ leads to a drastic reduction of the Doppler width $\delta\nu_D$ of the sample by a factor of $\frac{1}{2}(dE/E)^{1/2}$. For values of $dE = 1$ eV and $E = 60$ keV, which are typical of collinear laser spectroscopy, the initial Doppler width of about 1 GHz for the optical transition is reduced to values of $\delta\nu_D = 10$ MHz, comparable with the natural line width of strong optical transitions. The total Doppler shift is given by

$$\nu_D = \nu_0\ \sqrt{\frac{2\ E}{m\ c^2}} \tag{4}$$

Due to its mass dependence this leads to an "artificial" mass shift between two isotopes with mass m_1 and m_2 of

$$\delta\ \nu_D = \frac{\nu_0}{c}\ \sqrt{2\ e\ U}\ \left(\frac{1}{\sqrt{m_1}} - \frac{1}{\sqrt{m_2}}\right) \tag{5}$$

In the case of strontium isotopes this shift amounts to about 5.5 GHz for neighbouring isotopes in a typical optical transition of about 700 nm. Inserting these values into formula (1) leads to a single-step selectivity on the order of $S = 10^9$, which together with the mass resolution of the mass separator of about 1000 already meets the requirement as given above. Nevertheless the introduction of multi-step resonance ionization can further increase the selectivity by orders of magnitude.

IV. EXCITATION SCHEMES FOR RESONANCE IONIZATION OF STRONTIUM

In the experimental realization of resonance ionization of a fast atomic beam neither direct ionization nor excitation into autoionizing states can be performed due to strong background limitations in the collinear regime. A suitable procedure is found in the excitation of high-lying Rydberg states, which are subsequently field ionized in an electric field. As the oscillator strength of optical transitions connected with Rydberg levels varies in proportion to the cube of the main quantum number n and the critical field strength for the ionization process is proportional to n^{-4}, a compromise for the main quantum number of the Rydberg level chosen for excitation and field ionization has to be found. This value is in the range of $n \cong 15 - 20$. For the sake of simplicity the following discussion refers always to a level with main quantum number $n = 20$, which corresponds to a life time of the level of 3 µs and a critical field strength for ionization of about 4 kV/cm, which can be easily realized in the experiment.

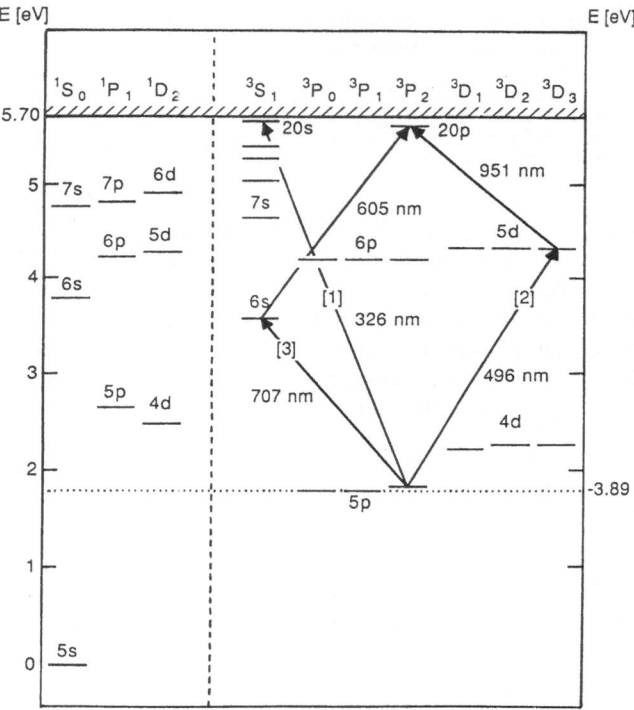

Fig. 2. Level schemes of strontium. At the right hand side
the ionization energy of the charge exchange medium
cesium is indicated. Three different excitation
schemes from the dominantly populated 5s5p $^3P_2$
level to Rydberg levels are given, with the wavelengths of each.

The level scheme of strontium is given in Figure 2. In collinear
fast-beam laser spectroscopy the strontium atoms are formed by neutralization of accelerated Sr^+ ions in a charge exchange process. The latter
has only a weak energy dependence. Thus during the neutralization, in
addition to the ground state, low-lying metastable states are also populated. As a consequence, almost no selectivity can be gained by laser
excitation of any ground state transition. In contrast, by choosing
cesium vapour with its low ionization energy of only 3.89 eV as the charge
exchange medium, dominant neutralization into the 5s5p states of the
triplet system can be achieved. Due to the statistical weigth of the metastable 5s5p $^3P_2$ level a population of about 20% in this state is expected.

Starting from this level the number of optical excitation steps by
visible light up to the ionization limit can be reduced by one. The
Rydberg state can be easily reached via the three different excitation
schemes indicated in Figure 2. Scheme (1) is the direct single photon
transition 5s5p $^3P_2$ - 5s20s $^3S_1$ at 326 nm. It has the advantage of
involving only a single laser step, while schemes (2) and (3) both involve
two-step excitations into the 5s20p $^3P_2$ via two different intermediate

168

states. Scheme (2) utilizes the first excitation step into the 5s5d $^3D_3$ level at 496 nm and then a second excitation at 951 nm, while scheme (3) involves a 707 nm transition into the 5s6s $^3S_1$ state and a further 605 nm excitation. Scheme (3) is hampered by strong optical pumping via the fast decay of the $^3S_1$ state into the $^3P_{1,0}$ levels. Thus efficient excitation in this scheme can only be performed by using pulsed lasers. This is not at all desirable due to the strong reduction in sensitivity caused by the low duty cycle of these systems. Even the use of high repetition systems such as Cu vapour laser pumped dye lasers with repetition rates of up to 10 kHz, and assuming perfect saturation of the optical transitions, leads to reduction factors of 50. For these reasons only schemes (1) and (2) will be discussed in detail.

V. EFFICIENCY OF THE METHOD FOR POSSIBLE EXCITATION SCHEMES

When applying cw lasers, the main limiting factor for the overall efficiency of the method is the efficiency of the optical excitations. In saturation of the optical transitions, a maximum efficiency of 50% per step can be reached for ideal geometrical and frequency overlap of the laser with the fast atomic beam. Unfortunately, the necessary laser power cannot yet be easily achieved for all frequency ranges with the present-day single-mode cw dye lasers. Only a particularly low excitation efficiency can be presently realized for the extremely narrow direct excitation line into the Rydberg state at 326 nm. In general the saturation power of an optical transition can be estimated[16]

$$\Phi \cdot \sigma \cdot t = 1 \tag{6}$$

Here Φ is the photon flux per unit area. The optical cross section σ as a function of the frequency ν is given by

$$\sigma\ (\nu) = \frac{\lambda^2}{2\pi} \frac{\Gamma_p}{\Gamma_t} \frac{\frac{1}{4}\Gamma_t^2}{(\nu_0 - \nu)^2 + \frac{1}{4}\Gamma_t^2} \frac{g_\beta}{g_\alpha} \tag{7}$$

with the wavelength λ, Γ_p and Γ_t being the partial and total radiative width and g_α and g_β the statistical weights of ground and excited state. The excitation time is limited to about 3 μs by the speed of the atoms and the excitation length of 1 m. With these parameters the results collected in Table 1 are obtained. For completeness the selectivities of the different transitions are also included in the compilation. The comparison shows that scheme (2) is the most favourable for the application envisioned.

In addition to the excitation efficiency the value for the overall efficiency of the complete experimental system must also account for sample preparation, production of the fast atomic beam and detection of the photo ions after excitation. The latter can be performed by low-background counting involving particle multipliers with an efficiency close to unity. The production of the fast atomic beam is much more complicated: it consists of the ionization process of strontium in the ion

Table 1. Selectivities and efficiencies of the optical excitation schemes discussed in the text. All values for the selectivity are given for the pair ^{90}Sr - ^{88}Sr.

Scheme Transition(s)	Wavelength (nm)	Selectivity at 60 keV	Power for Saturation (mW/cm^2)	Efficiency
(1) 5s5p $^3P_2$-5s20s $^3S_1$	326	3×10^9	200	3%[a]
(2) 5s5p $^3P_2$-5s5d $^3D_3$	496	2×10^6	0.24	50%
5s5d $^3D_3$-5s20p $^3P_2$	951[b]	6×10^5	5.1	50%
(3)[c] 5s5p $^3P_2$-5s6s $^3S_1$	707	7×10^5	***	0.8%
5s6s $^3S_1$-5s20p $^3P_2$	605	1×10^6	***	

(a) Value for realistic laser power of 5 mW/cm^2
(b) Frequency not yet easily realizable with single mode lasers
(c) Estimate for Cu-vapour pumped pulsed laser system.

source, having a probability of about 30% - 50%, the transmission of the mass separator and subsequent laser spectroscopic experiment ($\approx$50%) and the probability for charge exchange into the metastable 5s5p $^3P_2$ state ($\approx$20%). The efficiency of the chemical separation of strontium can reach about 50%. Thus the value for the overall efficiency is obtained by multiplying the excitation efficiencies of Table 1, yielding a factor of about 10^{-2}.

VI. DETAILS OF THE EXPERIMENTAL SET-UP

The complete measuring procedure of an environmental sample is composed of up to three different preparation steps and the final step of resonance ionization spectroscopy.

VI.1. First Preparation Step: Collection and Chemical Separation

The collection of air samples can be carried out by utilizing standard air filters. For the chemical separation of strontium a standard procedure can be used, applying different precipitation steps for the removal of Ca and Ba from the sample. As an alternative, crown ethers can be applied. Strontium and barium are separated from calcium by liquid-liquid extraction with dicyclohexyl-18-crown-6. After back-extraction of both elements from the organic phase, barium is extracted into a solution of dibenzo-24-crown-8 in nitrobenzene. Finally, strontium is deposited on a tungsten filament by molecular plating.

VI.2. Second Preparation Step: Pre-enrichment

For the detection of minimum concentrations of ^{89,90}Sr, samples from up to 1000 m^3 air must be processed, leading to quantities of stable strontium of $\simeq$ 1 mg. This corresponds to about 10^{19} atoms, which cannot easily be handled in the mass separator nor in the resonance ionization apparatus, for an acceptable measuring time. Thus pre-enrichment in the existing high-current mass separator facility SIDONIE 2 at GSI is foreseen for such large samples. The process leads to an enrichment factor of at least 1000 and has efficiencies of better than 10%, reducing the sample to 10^{14} atoms[17].

VI.3. Main Step: Resonance Ionization in a Fast Atomic Beam

The tungsten foils containing the chemically extracted or pre-enriched strontrium are directly introduced into the surface ionization source of the mass separator. At a temperature of about 2000 °K the strontium is quantitatively evaporated and ionized on a hot tungsten surface with ionization efficiencies of at least 30% - 50%. The extraction of the resulting ion beam is carried out by a 60 kV two-stage acceleration gap, and beam focusing is achieved by a set of one einzel and one quadrupole lens, producing a source emittance perfectly matched with the 60° acceptance angle of the mass separator magnet. The mass separator has been designed to produce a mass resolution of about m / Δm > 2000 and a beam emittance of $\varepsilon \leq$ 3 mm mrad, required for good overlap of the atomic and the laser beam in the collinear fast beam laser set-up. It is equiped with a set of beam scanners and Faraday cups for permanent monitoring of beam shape and positioning, and a number of correction elements for beam steering.

The collinear fast-beam laser facility, as shown in Figure 3, is based on the standard design, as described e.g. in Refs. 11, 18. After deflecting the ion beam by 10° to overlap the laser beam, it is neutralized in the charge exchange cell. At vapour pressure of about 10^{-3} mbar quasi-resonant charge exchange with the cesium vapour is achieved predominantly into the 5s5p ^{3}P$_{0,1,2}$ states of strontium. For laser adjustement and stabilization an optical observation region is installed downstream, where the optical resonances of the stable strontium isotopes can be recorded by fluorescence detection. Directly following is a field free excitation chamber of 1 m length, where the isotope under investigation is excited into Rydberg states.

The Rydberg atoms then travel into the ionization region, where field ionization is carried out in an over-critical electric field of up to 10 kV/cm. The ionization field is longitudinal to the beam, giving rise to no deflection of the photo-ionized particles. On the contrary, the photo ions are accelerated and thus energetically marked with respect to nearly all collisionally produced ions in the beam path[19]. An energy selective detection of these ions is realized by a two step deflection of about 43°, focusing only the photo ions with the correct beam energy onto a particle multiplier. Additional cleaning of the atomic beam of any remaining ions or collision-produced Rydberg atoms formed in the charge exchange process

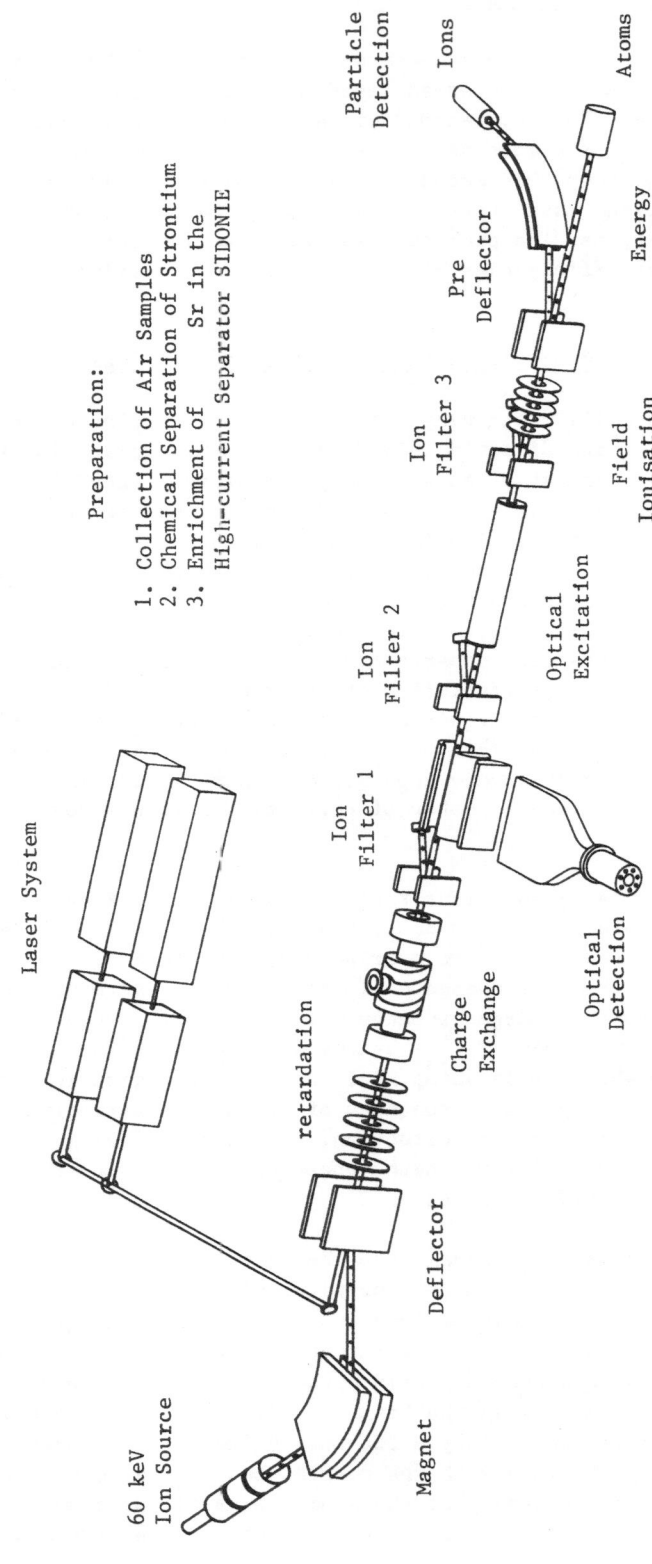

Fig. 3. Simplified sketch of the experimental set-up.

Laser System

60 keV
Ion Source

Magnet

Deflector

retardation

Ion
Filter 1

Charge
Exchange

Optical
Detection

Ion
Filter 2

Optical
Excitation

Field
Ionisation

Ion
Filter 3

Pre
Deflector

Energy
Selection

Particle
Detection

Ions

Atoms

Preparation:

1. Collection of Air Samples
2. Chemical Separation of Strontium
3. Enrichment of Sr in the
   High-current Separator SIDONIE

Table 2. Efficiency and selectivity in the detection of ^{90}Sr
with respect to the most aboundant stable isotope ^{88}Sr.

Step	Efficiency	^{88}Sr	^{90}Sr	Comments
Environmental Sample		10^{19}	10^8	after chemical separation
Enrichment at the high-current separator	10%	10^{15}	10^7	enrichment: 10^3 duration: 10^3 s
Production of the 60 kV ion beam	10%	10^{14}	10^6	surface ionization duration: 10^3 s
Mass separation	100%	10^{11}	10^6	enrichment: 10^3
Neutralization into the metastable level	10%	10^{10}	10^5	with deflection of residual ions
Optical excitation into the Rydberg level	> 3% >10^{-9}	10	3×10^3	resonant for ^{90}Sr
Field ionization and ion detection	30%	3	1000	50% efficiency 60% transmission

is performed by a set of three filter deflectors, located along the beam path directly after the charge exchange cell and at the entrance and exit of the excitation region. To permit the passage of the photo induced Rydberg atoms the voltage of the last deflector has to be adjusted very carefully just below the critical field value. It is believed that the use of this system under ultra-high vacuum conditions will reduce the background from collision-produced ions to a count rate of less than 100 counts for 10^{11} incoming projectiles. In this way nearly background-free counting conditions for ^{89,90}Sr can be realized.

After testing the overall efficiency of this set-up for all stable strontium isotopes and, in particular, calibrating the ratio of count rates for the different odd (^{87}Sr) and even (^{84,86,88}Sr) stable isotopes, the actual measuring sequence for the radioactive isotopes will be fixed. To avoid the complicated determination of all absolute efficiencies during each run, a known amount of a radioactive tracer material (probably ^{85}Sr) will be used for measuring relative count rates. For this purpose a fast cyclic switching of the excitation frequencies between the isotope under study and the tracer will be performed. This procedure is particularly simple in the case of the single photon excitation scheme (1), where it can be carried out via the Doppler effect by a well controlled variation of the beam energy, without changing the fixed laser frequency. In the case of two-photon excitation, in addition, the frequency of at last one laser has to be changed according to the isotope shift of the corresponding transition. This tuning procedure has to be carried out with a precision of about the experimental line width and thus will imply the absolute measurement and knowledge of the laser frequency to an accuracy of about 10 MHz.

VII. CONCLUSION AND OUTLOOK

The design of an experimental set-up is given for the detection of trace amounts of ^{89,90}Sr down to concentrations of about 10^8 atoms in an environmental sample containing about 10^{19}atoms of stable strontium. The apparatus is based on the technique of resonance ionization spectroscopy in collinear geometry, directly using the accelerated beam of fast atoms delivered from a mass separator facility after charge exchange. The set-up is presently under construction and will be ready for first measurements in 1990. In place of any final conclusions about the performance of the project we refer to Table 2, which gives a simplified scheme of the general procedure, separated into the processing of ^{90}Sr and the contaminating stable isotope ^{88}Sr. The comparison clearly demonstrates the overall concept and its performance.

VIII. ACKNOWLEDGEMENTS

This work has been funded by the German Federal Minister for Environment, Protection of Nature and Nuclear Safety.

REFERENCES

1. R. V. Abartsumyan and V. S. Letokov, Appl. Opt., 11:354 (1972)
2. V. S. Letokhov, "Laser Photoionization Spectroscopy", Academic Press, Orlando (1987)
3. G. S. Hurst and C. Grey Morgan, (eds.), "Resonance Ionization Spectroscopy", Proceedings of the Third International Symposium on Resonance Ionization Spectroscopy and its Applications, Swansea, Wales, Institute of Physics Conference Series 84, IOP Publishing Ltd., Bristol (1987)
4. T. B. Lucatorto and J. E. Parks, (eds.), "Resonance Ionization Spectroscopy", Proceedings of the Fourth International Symposium on Resonance Ionization Spectroscopy and its Applications, Swansea, Wales, Institute of Physics Conference Series 94, IOP Publishing Ltd., Bristol (1989)
5. R. D. Wilken and R. Diehl, Radiochimica Acta, 41:157 (1987)
6. H. O. Denschlag, A. Diel, K.-H. Gläsel, R. Heimann, N. Kaffrell, U. Knitz, K. Menke, N. Trautmann and M. Weber, Radiochimica Acta, 41:163 (1987)
7. H. Hotzl, G. Rosner and R. Winkler, Radiochimica Acta, 41:181 (1987)
8. T. B. Lucatorto. C. W. Clark and L. J. Moore, Opt. Comm., 48:406 (1984)
9. J. J. Snyder, T. B. Lucatorto and P. H. Debenham, J. Opt. Soc. Am., B2:1497 (1985)
10. S. L. Kaufman, Opt. Comm., 17:309 (1976)

11. R. Neugart, Collinear Laser Spectroscopy, in: "Progress in Atomic Spectroscopy" Part D, H. J. Beyer and H. Kleinpoppen, eds., Plenum Press, New York (1987)

12. E.-W. Otten, Nuclear Radii and Moment of Unstable Isotopes, in: "Treatise on Heavy Ion Physics", Vol. 8, D. A. Bromley, ed., Plenum Press, New York (1987)

13. S. V. Andreev, V. I. Mishin and V. S. Letokhov, Opt. Comm., 57:317 (1986)

14. Y. A. Kudriavtsev and V. S. Letokhov, Appl. Phys., B29:219 (1982)

15. Y. A. Kudriavtsev, V. S. Letokhov and V. V. Petrunin, Opt. Comm., 68:25 (1988)

16. W. Ruster, F. Ames, H.-J. Kluge, E.-W. Otten, D. Rehklau, F. Scheerer, G. Herrmann, C. Mühleck, J. Riegel, H. Rimke, P. Sattelberger and N. Trauttmann, Nucl. Instr. Meth., A281:547 (1989)

17. F. Nickel, K. H. Behr, A. Brünle and A. Steinhof, Nucl. Instr. Meth., B26:14 (1987)

18. A. C. Mueller, F. Buchinger, W. Klempt, E.-W. Otten, R. Neugart, C. Ekström and J. Heinemeier, Nucl. Phys., A403:234 (1983)

19. D. Dinger. J. Eberz, G. Hüber, H. Lochmann, G. Ulm and T. Kuhl, Z. Phys. D1:137, (1986)

LASER ULTRASENSITIVE PHOTOIONIZATION DETECTION OF RARE ELEMENTS

IN GEOCHEMISTRY, OCEANOLOGY AND STUDY OF GEOLOGICAL CATASTROPHES

G. I. Bekov and V. S. Letokhov

Institute of Spectroscopy
USSR Academy of Sciences
Troitzk, Moscow Region, USSR

I. INTRODUCTION

An investigation of the distribution of the rare elements in the
Earth's crust and in the ocean is of importance for many fields of
present-day natural science, including oceanology, geology, geochemistry
and cosmochemistry. Among such elements there are noble metals. The
research into their geochemical behavior, apart from its scientific
importance, is of practical use in search of their new promising sources
on land and sea.

Because the concentrations of noble metals in seawater are very low
(ppt level and less), their determination by classical spectral methods is
impossible without multi-stage concentration. The realization of these
approaches entails such difficulties that at this time there is scarcely
any informations available on the concentration of noble metals in
seawater. This applies equally to the measurement of noble metal concentrations in the Earth's crust, where the concentrations are one or two orders
of magnitude higher. For some platinum metals no data are available about
their content in many important geological rocks.

What restricts the sensitivity of traditional methods of analysis in
determining the content of noble metals in natural objects? The point is
that in the spectral methods of determination of element traces in a
substance most commonly used now, due to their relative simplicity, some
of the basic stages - atomization and detection of atoms - are realized
with an efficiency much less that unity. Such methods of optical detection as emission, absorption and fluorescence do not permit attaining the
ultimate detection level, that is, a single atom. Besides, in thermal
atomization of the substance in a buffer gas the atoms being detected may
undergo losses as a result of their high-temperature reactions with the
atoms and molecules formed by the atomization of the basic substance, and
with the impurities in the gas. The analysis becomes more complicated due
to so-called "matrix effects" when the analytical signal depends not only

Optoelectronics for Environmental Science, Edited by S. Martellucci and
A.N. Chester, Plenum Press, New York, 1990

on the concentration of the element to be detected, but also on the composition and form of the substence under test (or the matrix), and on the amount of third elements. The suppression of matrix effects is a principal problem in traditional analysis now. The new concentration techniques make it possible in some cases to increase the sensitivity of analysis by tens and even hundreds of times. However, in most cases the sensitivity gain does not exceed an order of magnitude.

It is clear that to solve the problem of detecting traces of noble metals in natural objects, it is necessary to apply methods with ultimate detection sensitivity and free of the above shortcomings. One of such methods developed in the last decade is laser stepwise photoionization, which under certain conditions provides registration of single atoms in a laser beam[1-3]. In combination with thermal atomization of samples in vacuum, this method was used by us for direct determination of impurities in high-purity materials, traces of elements in natural and biological objects[4,5]. It was shown that atomization in vacuum enables matrix effects to be suppressed to a considerable degree. The detection limits of elements obtained in direct analysis (without preliminary chemical preparation of the sample) range from 1 ppb to 10 ppt and in case of pre-concentration are 1 ppt and less[4].

Photoionization resonance spectroscopy, based on the multistep resonant excitation and subsequent ionization of excited atoms and molecules[1,2], is a unique method whereby all the ultimate measurement characteristics (sensitivity, spectral resolution, time resolution, selectivity, versatility) can be realized. For environmental studies the combination of ultra-high sensitivity and selectivity is most important.

This chapter presents a review of this method's analytical capabilities, based on the combination of atomic resonance photoionization with vacuum thermal vaporization and atomization of a sample, which allows detection of atomic traces at ppb-ppt level.

Examples of the use of ionization for ultrasensitive determination of element traces (Al, Au, platinum group elements, etc.) in real marine objects and traces of platinum group elements (Ru, Rh) in the Cretaceous-Tertiary boundary deposits demonstrates the extraordinary potential of this method for the environmental sciences.

II. EXPERIMENTAL TECHNIQUE OF LASER PHOTOIONIZATION ANALYSIS

In the method of laser stepwise photoionization the atoms are excited by laser radiation to an intermediate high-lying state through one or several steps, and then only the excited atoms are subjected to photoionization. There are three approaches used in this method, differing from each other in the manner of ionization of the atom fron an intermediate state. They are: i) nonresonant photoionization of the excited atom to the continuum; ii) resonant excitation of the atom to a Rydberg state with subsequent ionization by a pulsed electric field; iii) resonant photo-

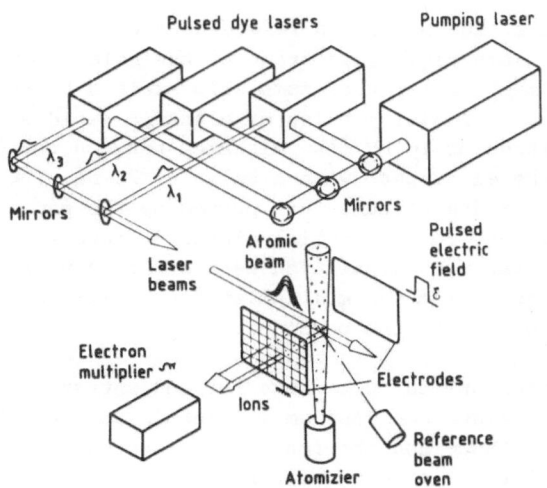

Fig. 1. Diagram of laser photoionization
analytical spectrometer.

ionization of excited atoms via a narrow autoionizing state lying in the
continuum.

As far as the use of laser stepwise photoionization for analytical
purpose is concerned, the most suitable approach for atom detection is
multistep excitation of atoms by laser radiation to a Rydberg state and
subsequent ionization of Rydberg atoms by an electric field pulse. In this
case the maximum elemental selectivity can be reached with minimal non-
selective ion background.

Figure 1 schematically shows a laser analytical atomic-photoioniza-
tion spectrometer (LAPIS). The laser part comprises a pumping laser (N_2-
laser, excimer laser or Nd:YAG laser) that simultaneously excites two or
three tunable dye lasers. This laser system makes possible the multistep
excitation of atomic Rydberg states for more than 80% of elements of the
periodic system.

In the process of analysis (Figure 1) the sample was placed into a
graphite, tantalum or tungsten crucible electrically heated to about
3000°K. The system of atomization and ion registration was located in a
vacuum chamber with a residual pressure of 10^{-6} Torr. In analysis of
solutions moisture was first evaporated from the crucible. The strong
background caused by thermal ions and electrons was removed by means of a
special ion suppression system installed above the heater.

The evaporating substance was formed by the crucible walls into an
atomic-molecular beam which passed between two electrodes while being
simultaneously irradiated at 90° angle by all the laser beams. For precise
tuning of the laser radiation to the excited transitions of the element
there was an additional furnace in the chamber which formed a reference

beam of this element. The basic and reference atomic beams crossed the
laser beams at the same region, in front of the hole in one electrode
covered by a wire mesh. After the atoms were excited by the laser pulses
to a Rydberg state, an ionizing pulse of electric field with its duration
of 10 ns and amplitude of 10 kV (the distance between the electrodes was
1 cm) was fed to the electrodes with a delay of 25 ns. The ions formed by
field ionization of Rydberg atoms were pushed by the same ionizing pulse
through the mesh in the zero potential electrode into a time-of-flight
mass filter, and after mass separation were directed with an electron
multiplier. The signal from the multiplier was amplified, supplied to the
boxcar averager and then to the recorder.

This method was applied to measure the concentration of aluminium,
ruthenium, gold, platinum and rhodium in oceanic samples. Both the direct
analysis of the substance and chemical sample decomposition followed by
preconcentration of the elements under study were used in these investigations. The experiments were performed with a laboratory laser photoionization spectrometer, and with an expeditionary version especially designed
for being installed on board a ship.

III. NOBLE METALS IN THE OCEAN

The behavior of noble metals in the ocean is the least explored field
of geochemistry. This is due to the fact that the content of these elements in seawater is extremely low and their analytical determination is
rather difficult. The required sensitivity of analysis must be from 10 ppb
to 0.1 ppt. Conventional methods enable the rather straightforward analysis
of sea ores and sediments for the content of Au, Pt and Ag. As for Ru, Rh
and Os, their content in the sea objects has not been really studied. The
most complicated problem, however, is to determine the content of noble
metals in seawaters. The LAPIS technique was applied to analyze a great
number of oceanic samples for Ru and Rh as well as to study the behaviour
of Au, Pt and Rh in hydrothermal springs of the Pacific.

Table 1 presents the result of determination of the concentrations
of a series of noble metals in various marine samples. The subjects of
investigation were mainly the hydrothermal springs of Juan de Fuca Ridge
and Guaymas Basin (Gulf of California). The samples included surface,
bottom and pore water. Bottom waters were sampled by means of a cassette
of "Rozett" barometers and "Pisces" submarine apparatus, and pore waters,
with a bottom scoop. The results for Au and Pt agree well with the data of
other authors[8,9]. As can be seen from the table, the rhodium concentrations are the lowest. This element is characterized by the greatest
scatter of concentration values. Taking into consideration the fact that
all samples were collected in areas showing hydrothermal activity, it can
be assumed that so great a scatter is due precisely to this activity. The
detection limits of the spectrometer in analyzing seawater (1 1) were
0.2 ppt for Au and 0.02 ppt for Pt and Rh (the respective detection limits
of the laboratory spectrometer version are about a factor of twenty
lower).

Table 1. Au, Pt and Rh concentrations in various marine samples from the Pacific. The values are given with an accuracy of ± 20%[6,7].

Element	Seawater, ppt			Suspension		Sulfide deposit, ppb
	Surface, 0-100 m	Bottom 5-300 m	Pore	"Black Smoker" ppb	Back- ground ppt	
Au	1-4	0.4-5.5	6-9	120	4	145
Pt	0.5-1	0.1-2.5	2-4.5	45	2	25
Rh	0.1-04	0.004-0.9	0.9-1.5	25	0.5	12

IV. CATASTROPHES IN THE GEOLOGICAL HISTORY OF THE EARTH

Extensive informations has been amassed in recent years about extraordinary events of very short duration that occurred more than once in the geological history of the planet Earth. Hypotheses have been advanced that those events caused by large extraterrestrial bodies that fell upon the Earth and left traces in the form of geochemical and isotopic anomalies which coincided in time with the major biostratigraphic boundaries[10]. The last of such events took place at the Cretaceous/Tertiary boundary more than 65 Myr ago. Alvarez et al.[10] took interest in the Great Mesozoic Extinction and decided to evaluate the rate of accumulation of sediments at this boundary by studying the concentration in sedimentary rocks of the platinum-group element iridium, which comes mainly from cosmic material. This element is a convenient indicator because extraterrestrial material is rich with it, whereas its concentration in surface terrestrial rocks is several orders of magnitude lower. The Ir concentration in the studied rocks turned out to be so high that it could not be explained by the gradual accumulation of extraterrestrial material in the form of micrometeorities and cosmic dust. This anomaly was interpreted as a result of a large cosmic body falling upon the Earth, which caused mass extinction of the biological species dominant in the Mesozoic, dinosaurs included. The existing alternative models of this event try to explain the origin of the anomaly of a number of elements by their concentration in the course of sedimentations or as a result of volcanic activity. To gain an understanding of the nature of the Cretaceous/Tertiary event, it is important to establish the proportions of various elements and their isotopes in the boundary deposits of that age, especially the proportions of platinum-group elements whose concentrations are high in extraterrestrial material and low in terrestrial rocks.

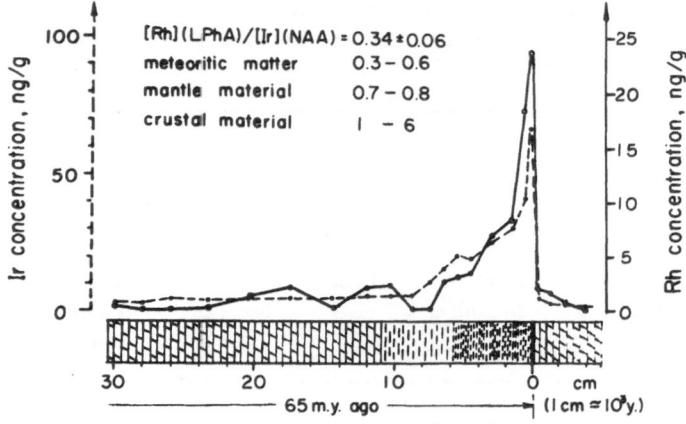

Fig. 2. Iridium and rhodium concentrations in the
Cretaceous/Tertiary boundary deposits
from the Sumbar-SM-4 section.

Unfortunately, it was impossible until very recently to establish
such proportions because of the lack of reliable methods for determining
platinum-group elements in rocks at a sensitivity level better than one
part per billion. It was only in some geological sections that the concentrations of platinum, palladium, ruthemium, osmium and its isotopes
were determined. The only exception was iridium, for it could be detected
in concentrations down to 10 ppt by the neutron-activation analysis technique. The extensive capabilities of the LAPIS technique have allowed
investigation of the proportions of platinum-group elements in the geological samples of interest. Figure 2 shows the distributions of iridium
and rhodium in the Cretaceous/Tertiary sedimentary rocks from the Sumbar-
SM-4 section in Turkmen S.S.R.[11]. The iridium content was determined by
neutron-activation analysis and the rhodium content by the LAPIS technique, the sample under analysis being subjected to chemical decomposition,
followed by sorption of the element of interest on a thio ether. It is
seen that the boundary between the geologic periods is clearly marked by a
sharp rise in the concentrations of both elements in the sedimentary
rocks. The Rh/Ir concentration ratio determined for this section correlate
well with that in meteorites. The data obtained point unambigously to an
extraterrestrial origin of the Cretaceous/Tertiary anomaly and confirm
the hypothesis of a large cosmic body colliding with the Earth. The
determination of the proportions of other platinum-group elements will
apparently allow the type of this body to be established.

V. ACKNOWLEDGEMENTS

The contribution of many collaborators including V. Radaev, E. Egorov
and M. Nazarov is gratefully acknowledged.

182

REFERENCES

1. V. S. Letokhov, in: "Tunable Laser and Applications", Springer Series in Optical Sciences, vol. 4, p. 122-139, Springer-Verlag, Berlin-Heidelberg-New York (1976)

2. G. S. Hurst, M. H. Nayfeh and J. P. Young, Phys. Res., A 15, 2283-2292 (1977)

3. G. I. Bekov, V. S. Letokhov, O. I. Matveev and V. I. Mishin, Optics. Letters, 3, 159-161 (1978)

4. G. I. Bekov and V. S. Letokhov, in: "Laser Analytical Spectro-chemistry", p.98-151, V. S. Letokhov, ed., Adam Hilger, Bristol (1986)

5. V. S. Letokhov, "Laser Photoionization Spectroscopy", Academic Press, New York (1987)

6. G. I. Bekov, V. S. Letokhov, V. N. Radaev, G. N. Baturin, A. S. Egorov, A. N. Kursky and V. A. Narseyev, Nature, 312, 748-750 (1984)

7. G. I. Bekov and A. S. Egorov, "Experimental Oceanography", Kluver, Holland (1990)

8. K. W. Bruland, "Chemical Oceanography", 8, 157-202, Academic Press, London (1983)

9. V. Hodge, M. Stallard, M. Koide and E. D. Goldberg, Anal. Chem., 58, 616 (1986); O. P. Svoeva et al., Zh. Anal. Chim., 40, 1606 (1985) (in Russian)

10. L. W. Alvarez, W. Alvarez, F. Asaro and H. Y. Michel, Science, 208, 1095 (1980)

11. G. I. Bekov, V. S. Letokhov, V. N. Radaev, D. D. Badykov and M. A. Nazarov, Nature, 332, 146-148 (1988)

DETERMINATION OF TOXIC METALS IN ENVIRONMENTAL OBJECTS

BY LASER EXCITED ATOMIC FLUORESCENCE SPECTROMETRY

M. A. Bolshov and V. Koloshnikov

Institute of Spectroscopy, USSR Academy of Sciences,
Troitzk, USSR

C. Boutron

Laboratoire de Glaciologie et Geophysique de l'Environnement
du CNRS, S. Martin d'Heres, France

C. Patterson

California Institute of Technology

N. Barkov

Arctic and Antarctic Research Institute, Leningrad, USSR

I. INTRODUCTION

For twenty years there has been a growing interest in the investigation of the occurence of Pb (and of several other heavy metals such as Cd,
Cu, Zn, Hg ...) in the well preserved dated snow and ice layers deposited
in the central areas of the remote Antarctic and Greenland ice sheets[1-5].
This is indeed a unique way to reconstruct the past natural tropospheric
fluxes of this highly toxic heavy metal on a global scale and to determine
to what extent these fluxes are now influenced by human activities.

Such investigation has unfortunately proved to be very difficult
because of the extremely low concentrations to be measured. As an
illustration[3], Pb concentrations in Holocene Antarctic ice have recently
been shown to be as low as about 0.4 pg/g. First of all, it is mandatory
to decontaminate the snow or ice samples before final analysis, since most
available samples are more or less contamined on their exterior, whatever
precautions are taken to collect them cleanly in the field[2,3,5]. Ultrasensitive analytical techniques must then be used. Due to the extremely
low concentration involved, ultraclean procedures are required throughout
the entire analytical procedure, from sample decontamination to final
analysis.

Up to now, the analytical techniques which have been used for the
final analysis of the decontaminated samples were either Isotope Dilution

Mass Spectroscopy (IDMS)[1-4] or Flameless Atomic Absorption Spectrometry (FAAS)[5,6]. The first of these two methods has the advantage of being absolute. Unfortunately, these two techniques do not allow a direct measurement of Pb at the extremely low concentration levels involved. For IDMS, a complicated chemical treatment, including an extraction into $CHCl_3$ and Dithizone, is required[1-3]. For FAASs, a preconcentration (by non-boiling evaporation or by absorption onto tungsten wire loops) is necessary[6]. These time consuming chemical treatments or preconcentrations are especially difficult to perform at the pg/g level. Moreover, rather large samples (30 to 500 g) are required in order to obtain high enough Pb amounts for the final analysis, and to keep these amounts higher than the contamination contributions from reagents and ware.

The newly developed technique of Laser Excited Atomic Fluorescence with Electro Thermal Atomization (LEAF-ETA)[7-9] offers a promising way to allow a direct and fast analysis of the decontaminated samples, using very small samples, without any preliminary chemical treatment or preconcentration. We present here the results of the first direct LEAF-ETA analyses of Pb down to sub pg/g concentration levels in Antarctic ancient ice using very small aliquots (only 20 µl) These results are compared with those previously obtained for the same ice samples by IDMS[3,4].

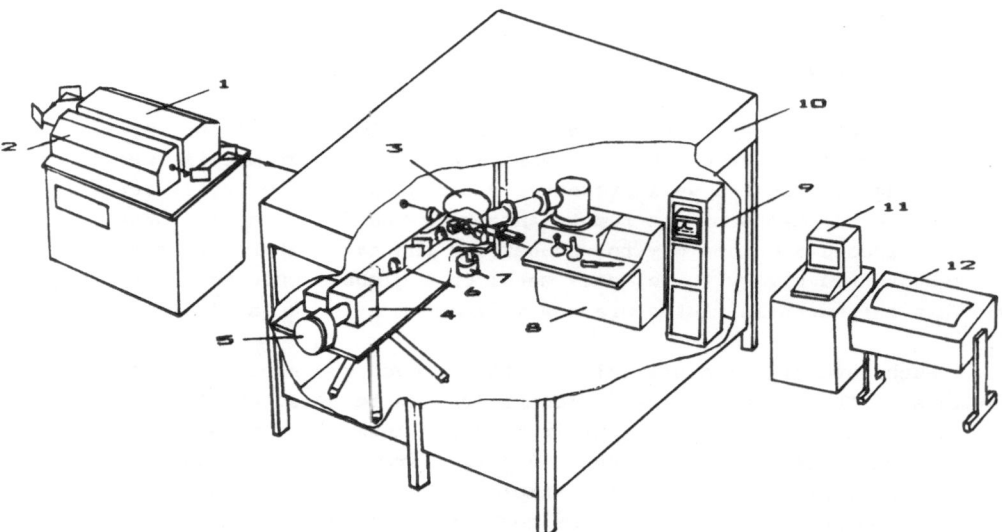

Fig. 1. Block diagram of LAFAS-1. (1) Excimer laser; (2) dye laser; (3) atomizer chamber; (4) monochromator; (5) PMT with pre-amplifier; (6) mechanical telescopic system with the lenses and filters; (7) temperature control system; (8) analytical table with vacuum pump system; (9) computer control system with KAMAK interface modules and TV set; (10) clean chamber; (11) terminal; (12) matrix printer.

In this lecture the results of the international collaboration of the scientists from France, USA and USSR are presented. The ice and snow sampling in Antartica was done mainly by Prof. C. Boutron and by Dr. N. Barkov. Samples preparation, decontamination and analysis by IDMS were done by Prof. C. Boutron and Prof. C. Patterson. The final analysis of the decontaminated samples by LEAFS-ETA technique were done by Drs. M. Bolshov and V. Koloshnikov and Prof. C. Boutron.

II. LASER EXCITED ATOMIC FLUORESCENCE SPECTROMETER

Recently, laser excited atomic fluorescence spectrometry has taken on a significant position among the sensitive analytical methods for determining trace impurities in solid, liquid and gaseous samples. The main advantages of this technique are: extremely high analytical sensitivity, wide linear range of the calibration curve, high selectivity of the analysis, comparative simplicity of the technique. When applied to pure aqueous standards, the LEAFS method, combined with sample atomization in an inert atmosphere in a carbon eletrothermal atomizer (ETA), provides elemental limits of detection (LOD) from 10^{-1} to 10^{-4} µg/l[8,9,11].

At the Institute of Spectroscopy of the USSR Academy of Sciences the foundations of the LEAFS-ETA method have been the object of investigations for some years and a laser excited atomic fluorescence spectrometer has been developed and modified. As a result of this work, a new version of the spectrometer laser atomic fluorescence automated spectrometer (LAFAS-1) has appeared. The analysis of Antarctic ice and snows were done at LAFAS-1. The block diagram of the spectrometer is shown in Figure 1.

The spectrometer comprises four basic units: a radiation source, an electrothermal atomizer, a recording system and a computer control system. Each of the four basic units is described in detail elsewhere[12]. We give here only brief descriptions of these units.

II.1. Radiation Source

The radiation source consists of a tunable dye laser (DL) pumped by XeCl excimer laser. The wavelength of the excimer laser radiation was 308 nm, the pulse energy was ut to 25 mJ, the repetition frequency was up to 25 Hz, the pulse duration was 10 ns, and the lateral dimensions of the laser output in the near field were 30 x 8 mm.

In the dye laser with transverse pumping a generator-amplifier scheme was used. The radiation energy of the excimer laser was split between the generator and the DL amplifier in the ratio 1:3.

The resonator of the DL generator was constructed from a grating (1200 lines/mm), a 6 - prism telescope (magnification 100), a dye cell (15 mm long) and an optical wedge (2°) as an output mirror. An original grating assembly and a control unit enable us to tune the DL wavelength using a microcomputer.

The generator radiation was amplified in the second cell (25 mm long) placed at 150 mm from the output mirror of the resonator. The coefficient of amplification varied from 10 to 50, depending on the energy of the excimer laser.

The spectral width of the amplified DL radiation was 0.8 cm^{-1}, and the efficiency of the excimer to DL energy conversion (at the wavelength 570 nm) was 9 - 10%.

Second harmonic generation (SHG) from the DL output was used to obtain the tunable UV-radiation. Three KDP crystals (40 mm length, 15 x 30 mm^2 cross-section) with the phasematching angles 54°, 59° and 64° provided SHG in the range 340 - 260 nm. To increase the SHG efficiency the DL radiation was focused into the crystal by a lens F = 110 mm. The construction of the SHG unit permitted us to tune the crystal angle both manually and by means of the computer controlled stepping motor. The radiation power of the second harmonic reached 3 kW.

II.2. Atomizer

In the spectrometer LAFAS-1 electrothermal atomization of the samples is used. The atomizer design allows one to make analyses both in the inert atmosphere and under vacuum. The chamber is made of a stainless steel tube with welded upper and bottom flanges. A lid and a base are hermetically fixed to the flanges by means of vacuum seals. The atomizer is attached to the base. It consists of two massive water cooled metal holders with graphite electrodes 6 mm in diameter fixed in them. A graphite cup is pressed between the graphite electrodes. Cups of different sizes were used: the outer diameter 5-6 mm, wall thickness 0.5-1.5 mm, height 4-7 mm. The cup can be quickly replaced trough the open lid. To replace the graphite electrodes it is necessary to lower the base.

The chamber has three branch pipes welded to it; they are provided with vacuum sealed quartz windows for the input and output of laser radiation and the output of fluorescence radiation. A special pipe connects the chamber with the pumping system consisting of a fore vacuum and diffusion oil-vapour pumps.

Sample atomization was carried out in five steps. At the first - evaporation - step the temperature could vary within 30-200°C during 10-120 s. At the second - ashing - step the temperature varied within 150-700°C during 1-60 s; at the third - atomization - step the temperature varied within 500-3000°C during 1-60 s; at the fourth - firing - step the temperature varied within 2000-3000°C during 1-10 s. The fifth - pause - step was used to cool down the cup to room temperature.

II.3. Recording System

The fluorescence radiation was collected from the analytical zone at a 90° angle and directed to the entrance slit of the monochromator by a telescopic system consisting of two identical quartz lenses with focal length F=150 mm and diameter 60 mm. The focal plane of the first lens coincided with the centre of the analytical zone, whereas the focal plane of the second lens coincided with the plane of the monochromator entrance slit. The spacing between the lenses was 100 mm. In this space neutral light filters could be placed to attenuate the fluorescence radiation when samples with high concentrations of the analyte were analyzed. The lenses and the filters were fixed in a special mechanical telescopic device connecting the chamber with the monochromator. This provided complete isolation of the monochromator entrance slit from light glare and scattered radiation in the room.

A compact diffraction monochromator ($260 \times 145 \times 135$ mm^3) with relative aperture 1:3.7 and reciprocal linear dispersion 6.3 nm/mm was used.

Behind the exit slit of the monochromator a photomultiplier tube (PMT) FEU-130 (or FEU-100) was placed. The PMT region of spectral sensitivity was 200-650 nm. The supply voltage of the PMT varied from 1.2 to 1.8 kV. The current pulse in the anode circuit of the PMT during the fluorescence pulse was amplified by a preamplifier (amplification factor 10) and fed to the input of the charge-sensitive (CS) ADC. The gate formation of the CS ADC was turned on by the leading edge of the pulse from the avalanche photodiode, to which a part of the excimer laser radiation was fed. The duration of the gate pulse was varied within 20-130 ns.

To reduce contamination problems, the whole spectrometer was located inside a specially designed room supplied with filtered air. Moreover, the electrothermal atomizer, the recording system and a small table covered with polyethylene film onto which the samples and standards were handled were placed inside a clean chamber ($2.5 \times 1.7 \times 2.8$ m^3). The ventilation system of the chamber provided high efficiency filtering of the air and laminar vertical flow of clean air inside the chamber.

Both the samples and the standards were introduced into the graphite cup of the atomizer using a 20 µl Eppendorf micropipette. The polypropylene tips of the micropipette were cleaned by dipping them for a few minutes into concentrated reagent grade HNO_3, then several successive rinsings with 1% HNO_3 (NBS twice distilled diluted in LGGE ultrapure water). The operator wore full clean room clothing and polyethylene gloves.

III. SAMPLES PREPARATION

The samples analyzed by LEAF-ETA in this work are fourteen sections of the 905 m Dome C deep Antarctic ice core   and six sections of the 2,083 m Vostok deep ice core[13]. Dome C core, which covers the past 40.000 years was thermally drilled at Dome C (77°39'S, 124°10'E) in a dry hole which had not been filled with a wall retaining fluid. The core from the Soviet Antarctic station Vostok (78°28'S, 106°48'E) was thermally drilled in a hole filled with a wall retaining fluid (kerosene). All these sections were decontaminated and prepared inside Patterson's clean laboratory at California Institute of Technology (Caltech, Division of Geological and Planetary Sciences), Pasadena, USA.

Each core section (diameter 10.5 cm, length 15-30 cm) was transported frozen from Antarctica to Caltech packed in double, sealed polyethylene bags. It was then decontaminated inside the Caltech clean laboratory by mechanically chiselling successive veneers of ice, progressing from the outside to the interior of the core using ultra-clean stainless steel chisels, inside a cooled double-walled nitrogen flushed ultra-clean polyethylene tray. The chiselling procedure has been described in details elsewhere[3]. The remaining inner core so obtained was 2 to 4 cm in diameter. Only these inner cores were analyzed by LEAF-ETA.

Each inner core was melted overnight at room temperature inside the Caltech clean laboratory in ultraclean 1 liter conventional polyethylene beakers. Twice distilled ultra pure HNO_3 prepared at US National Bureau of Standards (NBS) was then added to make a 0.1% HNO_3 solution. The Pb content of the batch of acid had previously been shown by IDMS to be 8 pg/ml[3,4]. The acidified solution was allowed to sit for two hours. A 50-100 ml aliquot was then taken into 250 ml conventional polyethylene bottles packed inside acid cleaned polyethylene bags. It was immediately refrozen and transported frozen to the Laboratoire de Glaciologie et Geophysique de l'Environment (LGGE), Saint Martin d'Hères, France.

Inside the LGGE clean laboratory these aliquots were melted at room temperature, and a 5-10 ml sub-aliquot was taken into 30 ml conventional polyethylene bottles packed inside acid cleaned polyethylene bags. These sub-aliquots were then transported frozen to the Institute of Spectroscopy in Troiztzk, USSR. Great precautions were taken during these successive aliquotings in order to insure an even distribution of the solid microparticles which might be present in the ice.

The 1 liter conventional polyethylene beakers and the 250 ml conventional polyethylene bottles had been cleaned inside the Caltech clean laboratory by a $CHCl_3$ rinse, at least three days at 55°C in 25% reagent grade HNO_3 (Baker), at least three days at 55°C in 0.1% HNO_3 (NBS twice distilled diluted in Caltech quartz distilled water (QDW), the Pb content of this water is 0.16 pg/ml), and finally at least three days in another 0.1% NBS HNO_3 bath at 55°C. The beakers were kept in the last bath for several months until use. The bottles were kept filled with 0.1% HNO_3 diluted in QDW for several months at room temperature before use.

The 30 ml conventional polyethylene bottles had been cleaned inside the LGGE clean laboratory using a procedure similar to that described in the previous section. The LGGE acid baths, however, were made from LGGE Maxy ultrapure water (the Pb content of this water has been shown by IDMS to be 0.27 pg/ml), and the bottles were allowed to sit in each of the three successive baths for at least two weeks.

Pb contamination introduced during the mechanical chiselling, from the NBS HNO_3, from the walls of the beakers, bottles and micropipettes and from the air in the clean laboratories was accurately determined from numerous blank determinations. For the 20 µl samples which will be used for the LEAF-ETA measurements, the corresponding Pb contamination value is found to be about 0.11 pg/ml, originating mainly from chiselling (about

0.1 pg/ml) and from HNO$_3$ (0.01 pg/ml). This represents from 0.5 to 30%
of total Pb present in these 20 l samples, depending on concentrations.
It must be emphasized that for the IDMS measurements of the same ice
samples[3], the relative contribution of the Pb contamination to the total
Pb present in the IDMS sample was higher (from 5 up to 75%)[3] because the
chemical treatment from sample melting to the final sample of isolated Pb
on the filament of the mass spectrometer is much more complicated than for
the LEAF-ETA measurements.

IV.   RESULTS AND DISCUSSION

IV.1. <u>Calibration of the Spectrometer</u>

    Detailed calibration was performed with synthetic acidified (0.1% NBS
HNO$_3$) multielemental standards. These standards simultaneously contained
19 elements, according to typical elemental concentration ratios of
Antarctic ice. They were prepared inside the LGGE clean laboratory from
Baker or Fisher certified atomic absorption standards (1000 ppm solutions). An intermediate solution containing all the 19 elements was first
prepared by diluting 10 to 500 µl of these standards in 200 ml of LGGE

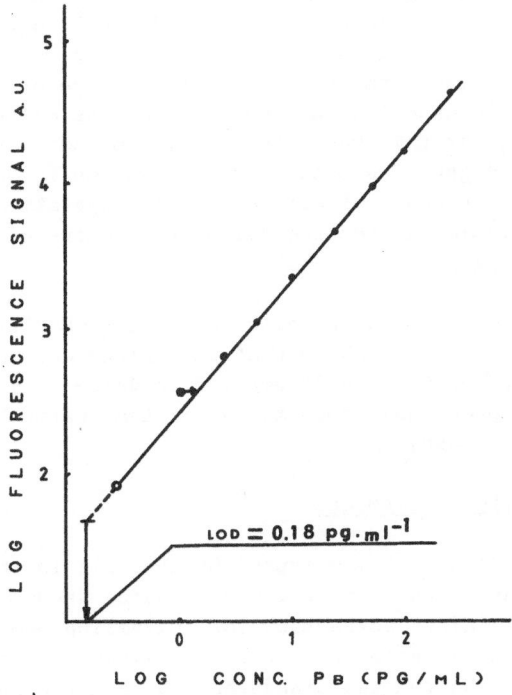

Fig. 2. Calibration curve for lead. 0 - lead
        concentration in the Grenoble ultrapure
        water. Arrow depicts the "true" value
        for 1 pg/ml standard.

ultrapure water in a 200 ml polypropylene volumetric flask. The low concentration standards were then made by diluting 2 to 1000 µl of these intermediate solution in 1000 ml of LGGE ultrapure water in a preconditionned ultraclean 1000 ml conventional polyethylene bottle. The standards were immediately acidified inside the bottle by adding 1 ml of NBS HNO$_3$ (measured in a specially designed FEP teflon graduated cylinder). The standards were then stored frozen in 30 ml preconditioned conventional polyethylene bottles packed inside acid cleaned sealed polyethylene bags. Cleaning procedures for the bottles and for the micropipette tips were similar to those described in the previous section. Pb concentration in the standards ranged from 1 to 500 pg/ml. Each of the standards was measured two to three times. The corresponding calibration curve is shown in Figure 2. The regression line was constructed using the least square method, in the form:

$$\log A_i = \log a + b \log c_i \tag{1}$$

where A_i is the ADC code proportional to the intensity of the fluorescence (arbitrary units), c_i is the Pb concentration in the ith standard, and b is the slope of the regression line.

When the calibration curve was constructed, the points 1 pg/g and 7.5 pg/g were not taken into account. The value measured for the 1 pg/g standard lies indeed significantly above the curve, because at this extremely low concentration level, the Pb content of the LGGE ultrapure water used for the preparation of the standards (0.27 pg/g) becomes significant and must be taken into account. The true Pb content in this 1 pg/g standard determined from the calibration curve of Figure 2 1.28 pg/g is indeed in excellent agreement with that calculated when adding the Pb content of this water to that introduced into the standard. The 7.5 pg/g standard gave systematically low values in all series of experiments, probably because of errors in its preparation (the Pb content in this standard calculated from the calibration curve of Figure 1 is 5.4 pg/g instead of 7.5 pg/g).

As an additional standard, we also analyzed LGGE ultrapure water three times, Figure 2. The Pb content of this water determined from the regression line is found to be 0.28 pg/g. This value is in excellent agreement with that previously determined for that water by IDMS (Boutron and Patterson, unpublished).

IV.2. <u>Dome C Antarctic Ice Samples</u>

Pb concentrations directly measured by LEAF-ETA in the inner cores of the fourteen sections of the Dome C ice core using sample volumes of 20 µl are shown in Table 1 (after correction for chiselling and nitric acid blanks, see Section III). Each inner core was measured only once. The calibration was based on the two standards 2.5 pg/g (which was measured two times) and 25 pg/g (three times).

The confidence intervals CI given in Table 1 were calculated using the equation:

$$CI = \frac{t_{P,f} \; S_{r,c} \; \overline{C}_x}{(N)^{1/2}} \qquad (2)$$

where $\overline{C}_x$ represents the mean measured Pb concentration in the sample, $t_{P,f}$ the students coefficient, N the number of parallel measurements of concentration C_x, $S_{r,c}$ the relative standard deviation of the measured values C_x, P the confidence level (P=0.95), m the number of points used to

Table 1. Comparative determination of Pb in the inner core of fourteen sections of the Dome C 905 m Antarctic ice core by LAFAS-1 and IDMS.

Sample depth (m)[a]	Measured Pb concentration[b] (pg/g)		
	LAFAS-1	CI[c]	IDMS[d]
172.8	1.44	0.7	0.76
300.6	0.69	0.45	0.47
373.9	0.84	0.5	0.94
451.9	0.32	0.2	0.43
476.3	0.16	0.2	0.32
500.5	4.6	2.2	3.81
527.2	6.8	3.1	10.5
545.1	7.3	3.3	10.2
602.2	9.9	4.4	11.4
658.2	13.8	6.1	14.0
670.5	19.7	8.8	29.3
704.2	13.4	5.9	15.2
775.7	6.2	2.8	7.2
796.9	1.84	1.0	1.25

(a) Age of the ice ranges from 3.800 to 34.000 years BP; (b) both LEAF-ETA and IDMS concentrations have been corrected for blanks; (c) confidence interval at the 0.95 confidence level calculated using equation (2), see text; (d) Pb concentration mesured in the same samples by IDMS.

construct the calibration curve, and f the number of degrees of freedom ($f=m-2$).

$S_{r,c}$ was calculated using equation 8 of Ref. 15.

It can be seen that the CI values given in Table 1 at the 95% confidence level are rather large: they range from about 30% up to 50% of the measured concentrations. These wide CI values are mainly due to the small number of measurements of the samples, the small amount of experimentally measured standards (m) and the small amount of background measurements. Indeed, we preferred analysing a larger number of ice samples with wide CI values, rather than analysing a small number of samples with smaller CI values. In this first use of LEAF-ETA for the direct measurement of Pb in Antarctic ice, the main goal was indeed to compare the results with those previously obtained by IDMS (see section IV.4) for a range of concentration as large as possible and for samples taken from different parts of the Dome C core (there might have been systematic differences since the speciation of Pb might not be the same during the Holocene (depths smaller than about 480 m) and during the Last Glacial Maximum (depths between about 480 m and 780 m).

The lowest Pb concentration values we measured with the spectrometer for the Dome C core were 0.27 pg/g (depth 476.3 m) and 0.43 pg/g (452.9 m). After substraction of the 0.11 pg/g blank (see section III), this gave final net concentration values of 0.16 and 0.32 pg/g, respectively, Table 1. The sample volume introduced into the spectrometer being only 20 µl, the absolute Pb mass which was detected in these two core sections was as low as about 5 to 8 fg (or ~ 1,5 to $2x10^7$ Pb atoms). This confirms the outstanding sensitivity of the LEAF-ETA technique, which is several orders

Table 2. Comparative determination of Pb in the inner core of six section of the Vostok 2.083 m ice core by LAFAS-1 and IDMS.

Depth[a] (m)	Estimated age (years BP)	Pb concentration (pg/g)	
		LAFAS-1	IDMS
489.90 - 499.20	26,200	42.4	38
851.45 - 851.70	53,600	4.5	3.1
1,425.11 - 1,425.51	99,700	2.7	2.4
1,775.32 - 1,775.54	129,000	2.6	2. .
1,850.35 - 1,850.54	134,000	13.7	10.6
2,026.22 - 2,026.36	155,000	19.1	19.8

(a) Depth are given as a real depth, they are not expressed as metres of ice equivalent.

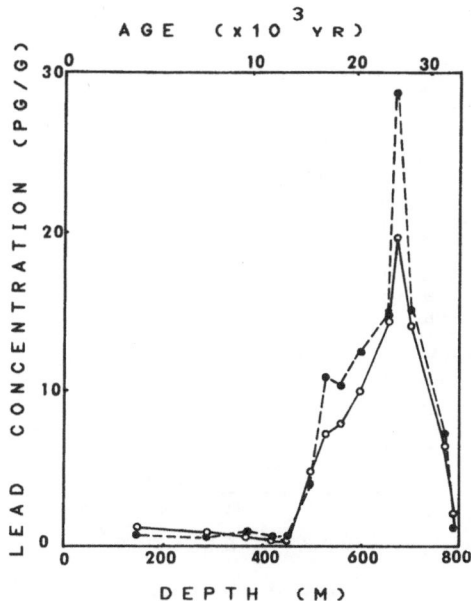

Fig. 3. Variations of Pb concentrations
in the inner parts of Dome C core.
LAFAS-1 results: (o, solid line);
IDMS-results: (●, dashed line).

of magnitude more sensitive for Pb than any other available analytical
technique.

IV.3. Vostok Ice Samples

Pb concentrations directly measured by LAFAS-1 in each of the six
Vostok inner cores (with duplicate determinations) are shown in Table 2.
Each concentration value has been corrected for the Pb contamination
introduced by the mechanical chiselling of the ice and by the HNO_3 added to
the samples as described above.

IV.4. Comparison with the IDMS Data

The inner cores of the Dome C and Vostok sample which we measured by
LAFAS-1 had previously been analyzed for Pb by the ultraclean IDMS technique in Patterson's laboratory[3]. The precision of these IDMS data, which
are shown in Tables 1 and 2, was estimated to range from about 50% (for
the very low concentrations at the 0.5 pg/level) down to about 10% (for
the high concentrations at the 5-30 pg/g level).

As shown in Tables 1 and 2 and Figures 3 and 4, LEAF-ETA and IDMS
data are in very good agreement, which strongly supports the accuracy of
both sets of data. It must however be emphasized again that for the LEAF-
ETA measurements, only 0.02 g samples were required even at the pg/g
concentration level, and that there was no need for any preconcentration
or chemical treatment: one sample determination took only 3-5 minutes. For

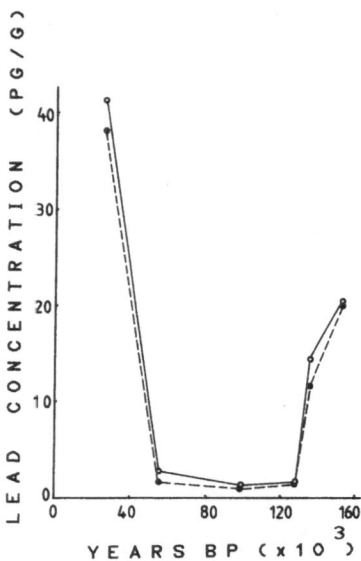

Fig. 4. Variations of Pb concentrations in
the inner parts of Vostok core.
LAFAS-1 results: (o, solid line);
IDMS results: (•, dashed line).

the IDMS measurements on the other hand, 50-200 g samples were required at
the pg/g concentration level, and a difficult and time consuming chemical
treatment was required: one sample determination took at least half a day.

The different sources of lead in Antarctic ancient ice were carefully
estimated and discussed in Refs. 3 and 4. It was shown[3,4] that the main
sources of atmospheric lead are soil-dust and volcano plumes. The lead
contribution from the oceans is insignificant throughout the period
studied. The time variations of lead faithfully track variations in soil-
dust lead, calculated from aluminum concentrations in the ice, thus
providing that the most of the natural lead in the global atmosphere
originated from soil dust, during both the late Wisconsin (Last Glacial
Maximum) and next to last ice age. The lead contribution from volcanoes,
calculated from sulphate concentrations in the ice, was negligible during
the late Wisconsin (LGM) and next to last ice age, but accounts for about
half of the measured total lead during the Holocene.

These data confirm that before humans began significantly to affect
the atmosphere, there was no excess of lead above that contributed by soil
dust and volcanoes, Now > 99% of the lead in the troposphere of the
Northern Hemisphere originates from human activities.

V. CONCLUSION

Our preliminary data confirm that LEAF-ETA is a very promising
technique for the direct measurement of Pb at extremely low concentration

levels in decontaminated samples of ancient Antarctic ice, using very small volumes of sample. However, it will be necessary in the future to improve the analytical precision by increasing the number of measurements both for the standards and for the samples.

It will be also be interesting to investigate the possibility of using LEAF-ETA for the direct measurement of other heavy metals such as Cd, Hg and Bi at very low concentration levels in Antarctic and Greenland ice and snow.

REFERENCES

1. M. Murozumi, T. J. Chow and C. C. Patterson, Geochim. and Cosmoch. Acta, 33, 1247 (1969)
2. A. Ng and C. C. Patterson, Geochim. and Cosmoch. Acta, 45, 2109 (1981)
3. C. F. Boutron and C. C. Patterson, Nature, 323, 222 (1986)
4. C. F. Boutron, C. C. Patterson, V. N. Petrov and N. I. Barkov, Atmospheric Environment, 21, 1197 (1987)
5. E. W. Wolff and D. A. Peel, Nature, 313, 535 (1985)
6. E. W. Wolff and D. A. Peel, Ann. Glaciol., 7, 61 (1985)
7. N. Omenetto and J. D. Winefordner, Laser atomic fluorescence spectrometry, in: "Analytical Laser Spectroscopy", N. Omenetto, ed., John Wiley, New York (1979)
8. M. A. Bolshov, in: "Laser Analytical Spectrochemistry", V. S. Letokhov, ed., Adam Hilger, Bristol (1986)
9. N. Omenetto, Spectrochim. Acta, 44B, 131 (1989)
10. M. A. Bolshov, C. F. Boutron and A. V. Zybin, Anal. Chem., 61, 1758, (1989)
11. M. A. Bolshov, A. V. Zybin and I. I. Smirenkina, Spectrochim. Acta, 36 B, 1143 (1981)
12. V. M. Apatin, B. V. Arkhangelskii, M. A. Bolshov, V. V. Ermolov, V. G. Koloshnikov, O. N. Kompanets, N. I. Kuznetsov, E. L. Mikhailov, V. S. Shishkovskii and C. F. Boutron, Spectrochim. Acta, 44 B, 253 (1989)
13. C. Lorius, L. Merlivat, J. Jouzel and M. Pourchet, Nature, 280, 644 (1979)
14. C. Lorius, J. Jouzel, C. Ritz, L. Merlivat, N. I. Barkov, Y. S. Korotkevich and V. M. Kotlyakov, Nature, 316, 591 (1985)
15. M. A. Bolshov, S. A. Dashin, A. V. Zybin, V. G. Koloshnikov, I. A. Majorov and I. I. Smirenkina, Zh. Anal. Chim., 41, 1862 (1986), (in Russian)

TRACE ANALYSIS OF PLUTONIUM AND TECHNETIUM BY RESONANCE IONIZATION

MASS SPECTROMETRY USING AN ATOMIC BEAM AND A LASER ION SOURCE

F. Ames, H.-J. Kluge, E.-W. Otten, B. M. Suri,* and
A. Venugopalan*

Institut für Physik, Johannes Gutenberg Universität
Mainz, Federal Republic of Germany

G. Herrmann, J. Riegel, H. Rimke, P. Sattelberger, and
N. Trauttmann

Institut für Kernchemie, Johannes Gutenberg Universität
Mainz, Federal Republic of Germany

R. Kirchner

GSI, Darmstadt, Federal Republic of Germany

I. INTRODUCTION

A method for low level detection of plutonium and technetium is described with a detection limit of less than 10^7 atoms. Plutonium is a very toxic element due to its radioactive decay as well as its chemical behaviour. It was released to the environment in large amounts during the fifties and sixties of his century, principally by nuclear-weapon tests and some accidents. As a result about 0.4 - 4 mBq per gram ^{239}Pu($T_{1/2}$ = 24390 y), corresponding to $4 \times 10^8 - 4 \times 10^9$ atoms, can be found in the Northern Hemisphere in soil samples.

Trace detection of plutonium has been done until now mostly by α-spectroscopy with a detection limit of about 0.4 mBq, requiring a measuring time of 1000 minutes. The origin of plutonium could be identified by determining the isotopic composition of the sample. This composition is difficult to determine by α-spectroscopy because the α-energies of ^{239}Pu and ^{240}Pu and of ^{238}Pu and ^{241}Am, a decay product of ^{241}Pu, are almost identical. The detection of other plutonium isotopes in the environment is handicapped either by their long lifetime, as in the case of ^{242}Pu, or by the absence of α-decay, as in the case of ^{241}Pu which is a pure β-emitter.

The experiments described here aim for a lower detection limit, which allows one to handle samples of smaller size. Furthermore, the measuring

---

*on leave from Bhabha Atomic Research Center, Bombay, India

time should be drastically reduced. Finally, it should be possible to measure the isotopic composition with a good accuracy.

Technetium is also important from the viewpoint of environmental studies, because $^{99}Tc(T_{1/2} = 2.1x10^5$ y) is a fission product of uranium. The oxide (Tc_2O_7) is quite volatile, and the water soluble pertechnetate ion TcO_4^- has high mobility in ground material. The isomer $^{99m}Tc(T_{1/2} = $ 6 h) is one of the most common isotopes in nuclear medicine. Hence a fast and sensitive method for trace analysis of technetium is required. Until now this has been done by β-spectroscopy, which requires a complicated chemical separation of the technetium to obtain a pure fraction, and leads to a detection limit of 0.05 Bq corresponding to $5x10^{11}$ atoms ^{99g}Tc.

Another important application for the sensitive detection of technetium is measurement of the solar-neutrino flux[1]. The isotopes ^{97}Tc and ^{98}Tc are formed by inverse beta decay, induced by solar neutrinos on molybdenum ore, by the reaction

$$^{97,98}Mo + \nu_e \longrightarrow ^{97,98}Tc + e^- \tag{1}$$

Due to its high threshold this reaction is sensitive only to the ^8B-neutrinos, which have also been measured by the Davis experiment for the last 20 years[2]. The results of this experiment disagree with the prediction of the standard solar model, which results in a two or three times higher value for the neutrino flux. In the case of the technetium solar-neutrino experiment, from the content of $^{97,98}Tc$ in molybdenum ore one determines the integrated solar neutrino-flux over an extremely long period. Such a molybdenum ore, with an age of about 4 million years, exists in the Henderson Rock Mine (Colorado, USA). A sample prepared by the chemical separation of Tc from 5000 t of this molybdenite is expected to contain about 10^8 atoms each of ^{97}Tc ($T_{1/2} = 2.6x10^6$ y) and ^{98}Tc ($T_{1/2} = 4.2x10^6$ y) and 10^{11} atoms of ^{99}Tc (fission product of uranium), along with $\cong 10^{15}$ atoms of Mo still remaining in the technetium sample. If the uranium content of the ore is known, which is the case for the Henderson Rock Mine, the amount of fission produced ^{99}Tc can be calculated exactly and used for calibration of the separation and detection efficiency of technetium. Hence it is only necessary to measure the isotopic ratios of the three technetium isotopes. The main requirements for such a geochemical neutrino experiment are extreme sensitivity and high elemental and isotopic selectivity. Radioactivity counting is not possible because of the low specific activity of the technetium isotopes. Mass spectrometry with positive technetium ions is excluded because of isobaric interference by molybdenum, and mass spectrometry with negative ions has failed until now[3].

Resonance ionization spectroscopy (RIS) and resonance ionization mass spectroscopy (RIMS) are promising techniques to fulfill the requirements listed above. The sample is evaporated and the free atoms are excited via several resonant transitions, and finally ionized. The last step can be accomplished by absorption of an additional photon either (i) non-resonantly by leading to the continuum, (ii) via a Rydberg state followed by field ionization, or (iii) by excitation of an autoionizing state. In the experiments described below a three colour, three step resonant excitation

via an autoionizing state has been chosen. Excellent elemental selectivity
is achieved by thus using three resonant steps. Furthermore, high sensitivity is obtained, as the cross section for excitation to an autoionizing
state can be larger by several orders of magnitude than that for a nonresonant excitation into the continuum.

II. RIMS ON A THERMAL ATOMIC BEAM EVAPORATED FROM A FILAMENT

II.1. Experimental Set-Up

The experimental set-up used has been described in detail in Ref. 4.
Therefore, only a few remarks will be made here. The set-up consists of a
dye-laser system used for resonant excitation and ionization, an atomic
beam source, a time of-flight (TOF) spectrometer and a data acquisition
unit for TOF and optical spectra. Three dye lasers (Lambda Physik, model
FL-2001) are pumped by one or two copper vapor lasers (Oxford Lasers,
model Cu40) with a pulse repetition rate of 6.5 kHz. The dye laser beams
are guided to the time-of-flight spectrometer through quartz fibres (d =
200 µm) and the out-coupled light is focussed into the interaction
region. The sample is deposited electrolytically on a rhenium filament
which is heated by a dc current so that the atoms under investigation
evaporate and form an atomic beam perpendicular to the laser beams. The
photoions created in resonance are accelerated by two grids to a final
energy of 2.9 keV, in such a way that a time focus is obtained at the end
of a two meter field free drift tube. Here the ions are detected by a
channelplate detector.

II.2. Measurements and Results

Several excitation schemes for plutonium and technetium were
tested[4,5] in order to select the one with maximum ionization efficiency
and hence sensitivity. Figure 1 shows the ion signal as a function of the

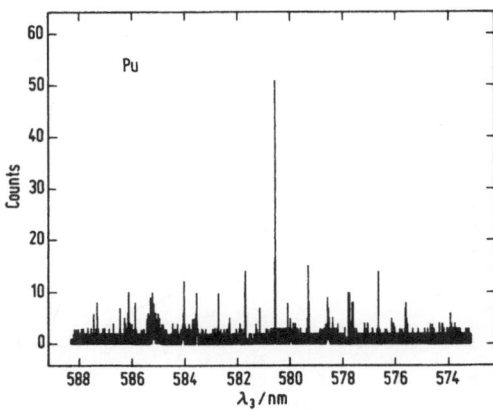

Fig. 1. Ion signal of plutonium as a function
of the third laser wavelength, with
λ_1 = 586.5 nm and λ_2 = 688.2 nm fixed
at resonance.

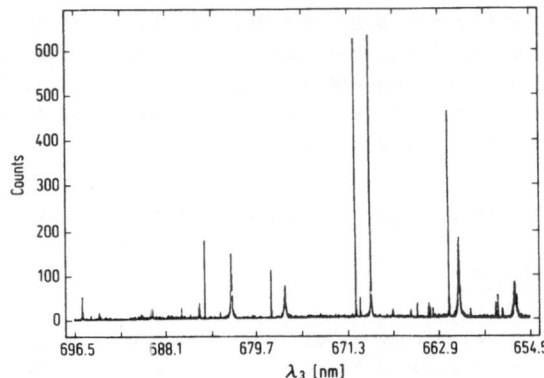

Fig. 2. Ion signal of technetium as a function
of the third laser wavelength, with
λ_1 = 313.1 nm and λ_2 = 821.1 nm fixed
at resonance.

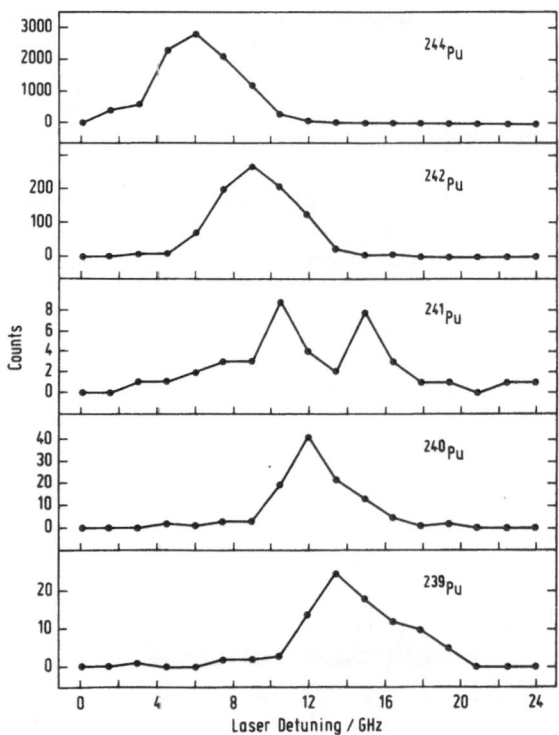

Fig. 3. Isotope shift and hyperfine structure in
the first transition of plutonium at
λ_1 = 586.5 nm.

third laser wavelength. Here the first two laser wavelengths are in resonance for the transition $5f^6 7s^2$ $^7F_1 \rightarrow 5f^6 7s$ $7p$ $^9G_1(\lambda_1 = 586.5$ nm) and $7s7p$ $^9G_1 \rightarrow 7s8s$ $^9F_1(\lambda_2 = 688.2$ nm). A strong increase in the ion signal was found by exciting the autoionizing state located 17220 cm$^{-1}(\lambda_3 = 580.5$ nm) above the $^9F_1$ state. An even larger enhancement in the ion signal was obtained for technetium, as shown in Figure 2.

To measure the efficiency ε of the detection system, a known number of atoms (minimum 10^8) was deposited on a filament. The ratio of the photoions detected to the number of atoms evaporated from the filament yields the efficiency.

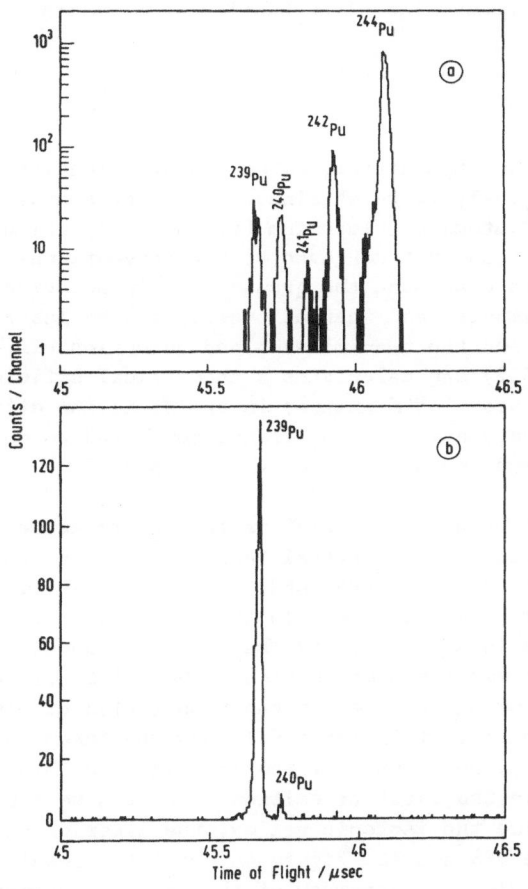

Fig. 4. Time-of-flight spectrum of a plutonium
sample. (a): Lasers scanned continuously
in the first and third transition to
cover the isotope shift and hyperfine
structure of the different plutonium
isotopes, (b): lasers set in resonance
for ^{239}Pu.

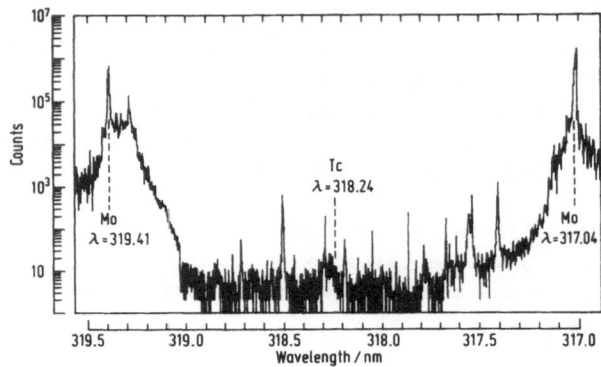

Fig. 5. Ion signal of molybdenum as a function of
the laser wavelength near the first optical
transition of technetium at λ_1 = 318.2 nm.

For plutonium and technetium, detection efficiencies of ε_{exp} = 4×10^{-6}
and 2×10^{-6}, respectively, were obtained. This corresponds to a detection
limit of around 10^6 atoms of plutonium (technetium), which is about 100
(10^5) times better than that achieved by the conventional α- (β-) counting
technique. Taking into account the ground-state population (0.6 for Tc
at 2300 °K), the temporal and spatial overlap of the laser beams and the
atomic beam (10^{-4}) and the transmission and detection efficiency of the
TOF spectrometer (0.2) one calculates a theoretical efficiency of $\varepsilon_{theo} \cong$
$\cong 10^{-5}$. The missing order of magnitude in the detection efficiency may be
due to incomplete saturation of the transitions, and to molecular evaporation of plutonium and technetium from the filament.

Another important aspect of RIMS is the determination of isotopic
ratios. This is influenced by several factors: the resolution of the TOF
spectrometer, hyperfine structure (HFS) and isotope shift (IS) in the
optical transitions, magnetic field in the interaction region, and the
polarization of the laser light. (i) The resolving power of the time-of-
flight spectrometer was measured as $M/\Delta M$ = 2500. This is easily sufficient
to separate neighbouring isotopes in the mass region of plutonium and
technetium. (ii) The effect of HFS and IS must be taken into account, as
shown in Figure 3 by the example of the ion signal obtained as a function
of the wavelength in the first transition of plutonium for five isotopes.
In order to determine the isotopic ratios, the laser wavelengths have to
be scanned over the HFS and IS. Figure 4a shows the result of such a scan
for a plutonium sample. The accuracy of the isotopic composition obtained
is about 10% compared with data obtained by mass spectroscopy. The isotopic
selectivity of RIMS is demonstrated in Figure 5b, which shows the result
obtained from the same plutonium sample with all lasers set on resonance
for ^{239}Pu. (iii) The dependence of ionization efficiency on the polarization of the laser light is minimized by using unpolarized light. Care has
to be taken, especially in case of isotopic ratio measurements of isotopes
with and without HFS.

The solar neutrino experiment requires good elemental selectivity between technetium and molybdenum. To estimate the supression of molybdenum, a molybdenum filament was used. The ion signal was recorded while scanning the laser over the first excitation wavelength of technetium λ_1 = 318.24 nm. Molybdenum has two strong ground state transitions in this region which enhance two photon-transitions to autoionizing states. The result of this measurement is shown in Figure 5. Comparing the maximum signal of molybdenum with the molybdenum signal obtained at the technetium wavelength one obtains an elemental selectivity S = 6.5x10 . This is on the lower side, as we were not able to saturate the two-photon transitions in molybdenum.

III. RIMS IN A CAVITY: THE LASER ION SOURCE

III.1. Principle and Experimental Setup

As discussed above, the most important limiting factors for the detection efficiency are the spatial and temporal overlap of the atomic beam with the laser beams, even at the very high repetition rate of our laser systems. To improve this the sample is placed in a hot cavity, with a small hole in it to feed in the laser beams for RIS and to extract the photoions produced. In this case the atoms have the chance to pass the interaction region of the laser light several times before they escape out of the cavity[6,7]. A similar construction was used by Andreev et al.[8,9]for the investigation of unstable francium isotopes. The efficiency of such a set-up is determined by the rate of photoionization R_I, the rate of diffusion out of the hole as neutral atoms R_D and the rate of surface ionization R_S, which is a non-selective undesired process. The total efficiency is given by

$$\varepsilon = \frac{R_I}{R_I + R_D + R_S} \tag{2}$$

For a cylindrical cavity with a length l and a diameter D = 1 and neglecting surface ionization, which is justified for technetium in a tungsten cavity (ε_{sur} = 10^{-6} at 2400 °K), one obtains

$$\varepsilon = \frac{\nu_{rep}\, \varepsilon_{photo}}{\nu_{rep}\, \varepsilon_{photo} + \overline{\nu}/4l} \tag{3}$$

Here ν_{rep} is the pulse repetition rate of the laser, ε_{photo} the photoionization probability at resonance and $\overline{\nu}$ the mean thermal velocity of the atoms in the cavity. The mean free path λ of atoms in the cavity between two wall collisions was taken as λ = 1. As can be seen from Eq. (3) the efficiency does not depend on the size of the hole in the cavity. This is true as long as the diameter of the laser beam is equal to or larger than the diameter of the hole. Using Eq. (3) one obtains a total efficiency of ε_{theo} = 0.2 for ^{99}Tc at 2400 °K, with a pulse repetition frequency of the lasers ν_{rep} = 6.5 kHz, a photoionization probability ε_{photo} = 0.6 (corresponding to the thermal ground state population at complete saturation of excitation and ionization), a thermal velocity $\overline{\nu}$ = 650 ms^{-1}, and a cavity

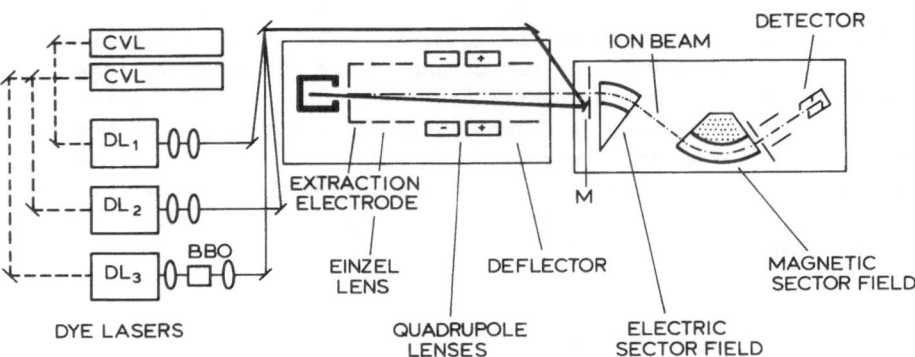

Fig. 6. Laser ion source for trace detection of technetium, with
the copper vapor laser system and the Mattauch-Herzog mass
spectrometer used for diagnostics.

length $l = 1$ cm. Hence an improvement in efficiency by several orders of
magnitude is expected by the use of a cavity instead of a filament.

The experimental set-up for testing this is shown in Figure 6. The
same laser system described earlier is used. The laser beams are deflected
via mirrors into the ionization cavity, which is a hollow tungsten
cylinder of inner length and diameter of $l = D = 1$ cm with two end cups.
One has a hole at the center with a diameter $d = 2.5$ mm. The cavity can be
heated by electron bombardment up to a maximum temperature of 2500 °K.
Such a design was developed at GSI, Darmastadt as a target ion source[10]
and was slightly modified for our requirements. The ion produced are
extracted by an extraction electrode, accelerated to 10 keV, and focused
by an einzel lens and a quadrupole doublet to the entrance slit of a
double-focussing Mattauch-Herzog mass spectrometer[11]. The use of a
magnetic rather than a TOF mass spectrometer is essential. This is because
of the high dc background of thermal ions, especially of alkalies due to
their low ionization energy, which are evaporated from the construction
material of the cavity and surface-ionized at the hot walls.

III.2. Measurements and Results

The initial measurements were performed with about 10^{14} ^{99}Tc atoms,
deposited electrolytically on a rhenium filament which was placed inside
the cavity. The cavity was heated to about 2400 °K. Figure 7a shows a mass
spectrum recorded with laser beams off. It gives the thermally-ionized
molybdenum isotopes which are present as impurities in the tungsten
material (100 ppm) and a peak at mass $A = 99$, which is thermally-ionized
technetium. The ratio of ^{99}Tc/^{98}Mo is $R = 1/4$. In Figure 7b, the signal
displayed is obtained from the same sample with lasers set on resonance
for photoionization of ^{99}Tc. The count-rate of ^{99}Tc ions increased by a
factor of 80 at an essentially vanishing background, and the ratio R has
changed to $R = 20$. As the lasers are pulsed the photoion beam is also
pulsed. Hence it is possible to gate the detection system in order to
suppress the thermal molybdenum ions which are continuously produced.

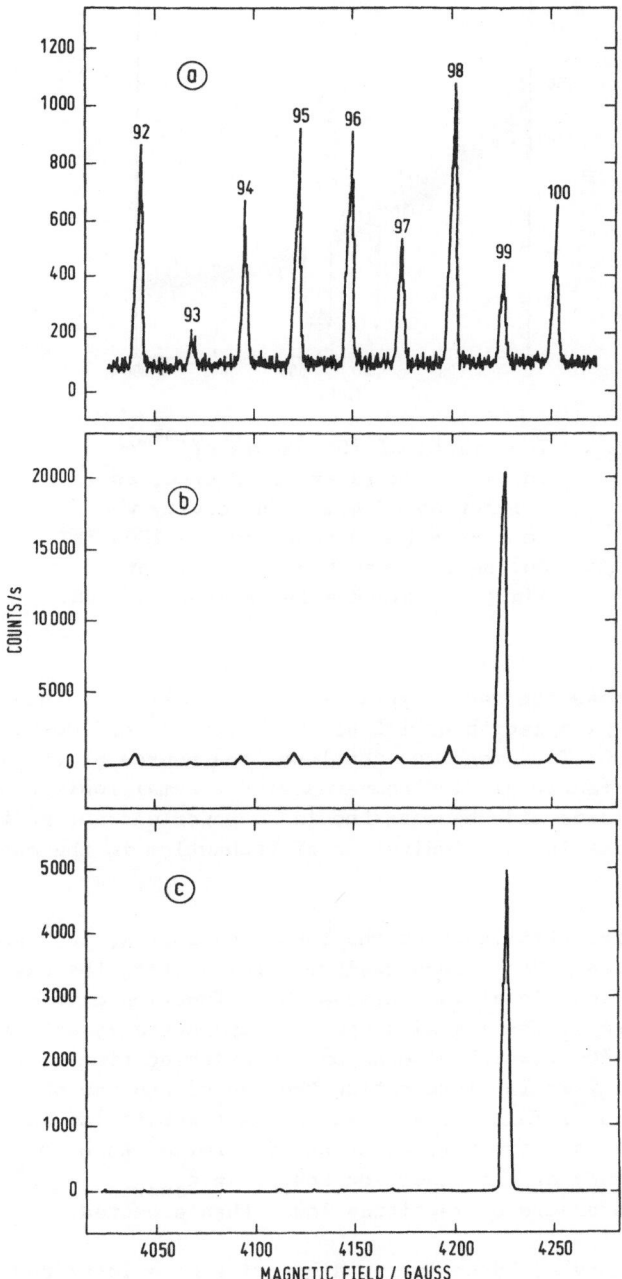

Fig. 7. Mass spectrum of a ^{99}Tc-sample obtained with laser ion source. (a) Laser beams off. (b) Lasers on and wavelengths set on resonance for photoionization of ^{99}Tc. (c) Same as (b) but with gated photoion detection.

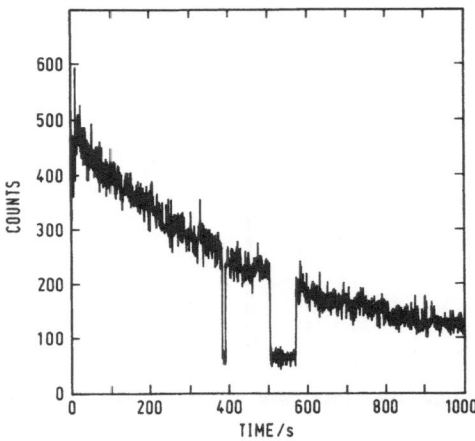

Fig. 8. Ion signal of 10^{11} atoms of ^{99}Tc
put into the laser ion source, as
a function of time. The cavity was
kept at a temperature of T = 2500 °K.
During the measurement the laser
light was blocked for a short period.

Figure 7c shows the same signal as Figure 7b but obtained by gated
detection with a gate length of 1.5 μs. The ratio ^{99}Tc/^{98}Mo has now
changed to R = 600. To summarize, the laser ion source provides a gain in
selectivity by a factor of 2400 compared with thermal ionization. This
value is on the lower side because the laser power was not sufficient to
saturate the excitation and ionization of technetium in the experiments
described.

To measure the efficiency of the laser ion source, 10^{11} atoms of
^{99}Tc were deposited into a clean cavity. After heating the cavity to
2500 °K the photoion signal was observed as a function of time. The result
is shown in Figure 8. The signal decreases exponentially with a time
constant of τ = 440 s, which depends on the sticking time of technetium at
the tungsten cavity walls. Integrating the ion counts one obtains a total
efficiency $\varepsilon = 2 \times 10^{-6}$. This value includes the transmission and detection
efficiency of the spectrometer, which was determined to be $\varepsilon_{spec} = 10^{-3}$.
Hence the efficiency of the laser ion source is $\varepsilon_{source} = 2 \times 10^{-3}$.
This value is two orders of magnitude lower than expected.

This efficiency could be improved by using more laser power, as the
excitation steps in the photoionization of technetium are not being
saturated with the laser intensity presently available. Furthermore,
because of the inhomogeneity of the beam profiles, the spatial overlap of
the laser beams is smaller than the size of extraction hole, resulting in
a reduction of the effective interaction volume and hence a lower efficiency.
This could be improved by the use of quartz fibers for the transmission of
the laser light to the laser ion source, because a more homogenous beam
profile is obtained in this way. Finally, the possibility of some of the

^{99}Tc-atoms remaining in the hot cavity even after prolonged heating cannot be ruled out. This needs to be checked in further experiments by β- or γ-spectroscopy.

IV. CONCLUSION

Resonance ionization mass spectroscopy is a powerful technique for trace analysis. This has been demonstrated in the case of plutonium and technetium employing a time-of-flight mass spectrometer. The overall efficiency is about 10^{-6}, resulting in a detection limit of less than 10^7 atoms in a sample. The efficiency could be considerably further improved by the use of a laser ion source, as has been shown in the case of technetium. An efficiency of better than 10^{-3} has been achieved, which however does not include the transmission and detection efficiency of the mass spectrometer used. In addition to the extreme sensitivity, high elemental and isotopic selectivity is obtained. However for the geochemical solar neutrino problem it is necessary to reduce the background caused by surface ionization of molybdenum at the cavity walls. Use of purer materials containing less molybdenum for construction of the laser ion source cavity could be a solution of this problem. It might also be solved by the use of wall material with a lower workfunction than that of tungsten (for example carbides) to suppress thermal ionization. Work in this direction is in progress.

V. ACKNOWLEDGEMENT

This work has been funded by the German Federal Ministery for Research and Technology (BMFT) and the Deutsche Forschungsgemeinschaft (DFG). We also wish to thank the Laboratoire René Bernas, Orsay, France for making the Mattauch-Herzog spectrometer available to us.

REFERENCES

1. G. A. Cowan and W. C. Haxton, Science, 216:51 (1982)
2. J. N. Bahcall and R. K. Ulrich, Rev. Mod. Phys., 60:297 (1988)
3. D. J. Rokop, N. C. Schroeder and K. Wolfsberg, High sensitive technetium analysis using negative thermal ionization mass spectrometry, in: "Advances in Mass Spectrometry", Proceedings of the 11th Int. Mass Spectrometry Conf. Bordeaux, 1988, P. Longevialle ed., Heyden and Son, London (1989)
4. W. Ruster, F. Ames, H.-J. Kluge, E.-W. Otten, D. Rehklau, F. Scheerer, G. Herrmann, C. Mühleck, J. Riegel, H. Rimke, P. Sattelberger and N. Trauttmann, Nucl. Instr. and Meth. A281:547 (1989)
5. P. Sattelberger, M. Mang, G. Herrmann, J. Riegel, H. Rimke, N. Trauttmann, F. Ames and H.-J. Kluge, Radiochim. Acta, 48:165 (1989)
6. H.-J. Kluge, F. Ames, W. Ruster and K. Wallmeroth, Laser Ion Sources, Proceedings on Accelerated Radioactive Beams Workshop, Vancouver

Island (1985) in L. Buchman J. M. D'Auria eds., TRIUMF Proceedings
TRI-85-1:119

7. S. V. Andreev, V. I. Mishin and V. S. Letokhov, Optics. Com., 57:317
   (1986)

8. S. V. Andreev, V. S. Letokhov and V. I. Mishin, Phys. Rev. Lett.,
   59:1274 (1987)

9. S. V. Andreev, V. I. Mishin and V. S. Letokhov, J. Opt. Soc. Am.,
   B5:2190 (1988)

10. R. Kirchner, K. H. Burkard, W. Hüller and O. Klepper, Nucl. Instr.
    and Meth., 186:295 (1981)

11. M. Epherre, G. Audi, C. Thibault, R. Klapisch, G. Huber, F. Touchard
    and H. Wollnik, Nucl. Phys., A340:1 (1980)

OTHER TRACE ELEMENTS

TUNABLE DIODE LASER SPECTROSCOPY FOR THE RECONSTRUCTION

OF ATMOSPHERIC CO_2-CONCENTRATIONS OF THE PAST 50,000 YEARS

B. E. Lehmann

Physics Institute
University of Bern
Bern, Switzerland

I. INTRODUCTION

The continuous measurements performed since 1958 at Mauna Loa
(Hawaii) show that the global atmospheric CO_2-concentration is increasing[1]. Since carbon dioxide is known to be an active greenhouse gas it is
to be expected that global climate changes will result. Improved numerical
climate models will eventually be able to predict the size and regional
pattern of these changes. In order to validate such models it is important
to know the history of the climate on Earth. Atmospheric air, which is
enclosed in bubbles in polar ice samples offers a unique possibility to
reconstruct the composition of the Earth's atmosphere and in particular to
determine how the CO_2-content has varied in the past. Several deep
drilling projects in Greenland and in Antarctica have ice cores made
available that represent snow fall and corresponding trace substances from
the atmosphere that reach up to 100,000 years and more back in time. Since
these samples are obviously not easy to get it is important to use analytical techniques that needs as little of the valuable ice as possible for a
measurement. For more than ten years now we have used the laser spectrometer[2,3] described in the following sections, and with its use we were
able to reconstruct the atmospheric CO_2 content during the past 50,000
years[4-7].

II. TUNABLE DIODE LASER SPECTROMETRY

The system is schematically shown in Figure 1. The main elements are
a PbS diode laser, a short gas cell for direct absorption measurements and
a PbSe detector for recording the transmitted light as the laser is tuned
over an absorption line of CO_2. Further elements in the light path include
a single lens for focussing the beam through the cell, a mechanical
chopper for modulating the beam in order to detect the signal by lock-in-
amplification, a monochromator used as a mode filter to separate the
individual laser modes and a separate small absorption cell which is occa-

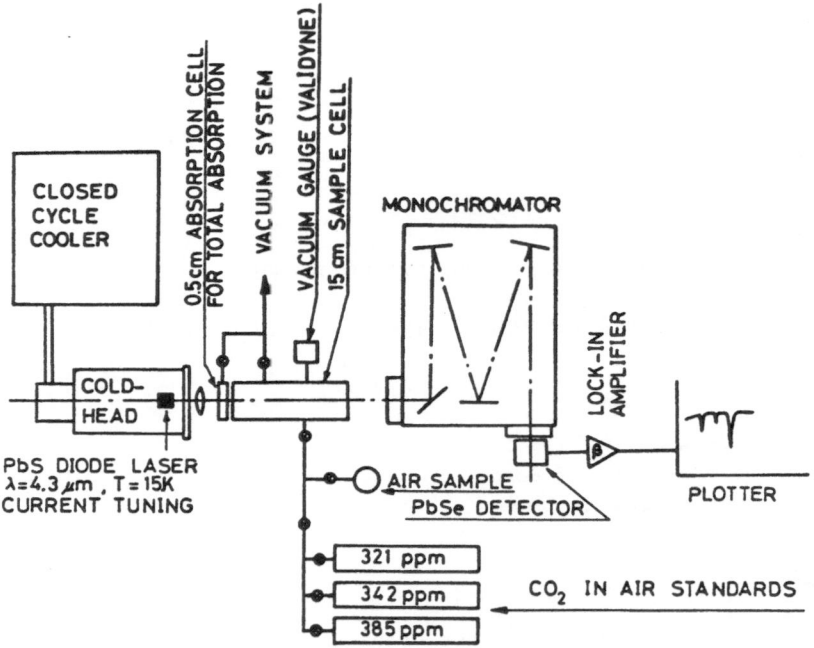

Fig. 1. Schematic representation of the Tunable Diode Laser
Spectrometer[2].

sionally filled with a few mbar of pure CO_2 for determining the remaining
transmission with 100% absorption at line center.

The diode laser is mounted on a cold finger cooled by a closed cycle
refrigerator. It emits light in the 4.3 μm band of CO_2. The wavelength is
tunable over approximately 45 cm^{-1}. A single mode can usually be scanned
over a small interval of 1-2 cm^{-1}. With the proper temperature stabilization and current regulation the laser can therefore be tuned over a single
rotation-vibration line of CO_2. For maximum absorption one of the
strongest transitions in the P- or R-branch of the 00^00-00^01 fundamental
band of CO_2 is selected. The laser bandwidth is on the order of 10^{-4} cm^{-1},
much smaller than the absorption line width, which is Doppler-limited at
the low pressure filled into the absorption cell. The technical data of
the optical system are summarized in Table 1. The entire light path must
be in a CO_2-free atmosphere, therefore all components are enclosed in a box
through which dry nitrogen gas is flushed.

While tuning the laser wavelength slowly (within approx. 50 seconds)
over an absorption line of CO_2) the output of the lock-in amplifier is
digitized at a rate of 500 Hz and the signal is then further processed by
a personal computer.

214

Table 1. Technical data of the optical system.

light source:	PbS-diode laser
operating temperature:	$T < 36°K$
current:	$I < 1.5$ A
spectral range:	2290 cm^{-1} - 2435 cm^{-1}
current tuning-rate:	approx. 45 cm^{-1}/A
temperature tuning-rate:	approx. 1.1 cm^{-1}/°K
temperature stabilization:	± 0.3 m°K/min
average power (cw):	approx. 100 μW (multi-mode)
bandwidth:	approx. 10^{-4} cm^{-1}
lens:	$d = 2.5$ cm, $f = 2.5$ cm
chopper:	$f = 180$ Hz (mechanical)
monochromator:	0.5 m Czerny-Turner
slits:	200 μm (used as a mode filter)
detector:	PbSe
lock-in amplifier:	Ithaco Dynatrac 319A

III. SAMPLE PROCESSING AND ANALYSIS

Polar ice contains approximately 100 cm^3 STP of gas per kilogram of ice. We currently work with ice cubes of 10 cm^3, although smaller samples, of e.g. only 1 cm^3, can easily be measured. The ice is cracked under vacuum by a pneumatically driven matrix of needles to release the trapped air. From 10 cm^3 of ice we extract a typical gas samples of 1 cm^3 STP. The sample is transferred through a cold trap (at -20 °C), which retains any water vapour, into the absorption cell. The total volume of the inlet system, including the cracker, the water vapour cold trap, a Barocell pressure gauge and the 30 cm-absorption cell is about 200 cm^3. A typical gas sample therefore produces a total pressure of 5 mbar in the absorption path, which represents a gas density of 1.3×10^{17} molecules/cm^3. At a concentration of 300 ppm the concentration of CO_2 is then 4×10^{13} molecules/cm^3 in the absorption cell.

Line center absorption coefficients under these conditions are on the order of 5×10^{-16} cm^2/molecule and an air sample with 300 ppm of CO_2

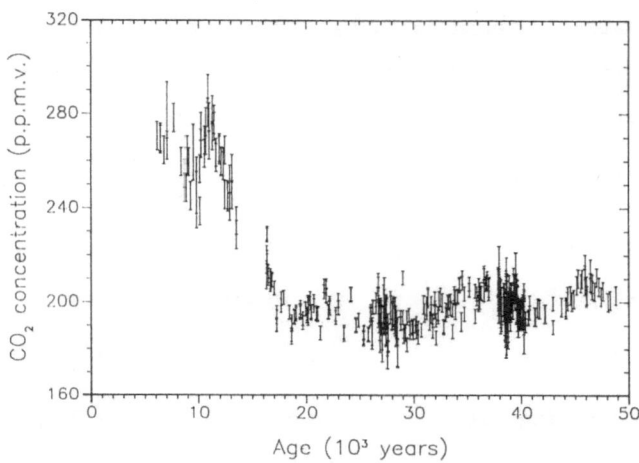

Fig. 2. Measured CO_2-concentrations
in the Byrd ice core[6].

yields a total absorption of almost 50% in our 30 cm absorption cell. We therefore can measure absorption peaks in a range where the signal is most sensitive to small variations in the concentration. Absolute calibration is achieved by comparing the absorption with that of a standard gas with e.g. 321.1 ppm of CO_2 in air of which an amount is released to the absorption cell that matches exactly the measured pressure of the sample. This calibration is made for each sample immediately after the gas has been pumped off.

The overall precision reached with this procedure is 0.25% (± 0.75 ppm in 300 ppm) and the total time for one measurement is about 15 minutes per sample. This includes the removal of the previous sample, the introduction of the new sample, pumping to 4×10^{-5} mbar, the cracking of the sample and its transfer to the absorption cell, a total of 5 slow scans over the absorption line, the pumping of the cell and the filling and measurement of the standard gas and the data analysis.

IV. EXAMPLES OF RESULTS OBTAINED
WITH THE TUNABLE DIODE LASER SPECTROMETER

Several hundred samplese have been analyzed with the system during the past few years. As an illustration the results of the Antarctic ice core[6] from Byrd station are reproduced in Figure 2.

V. ACKNOWLEDGEMENTS

Members of the ice research group in Bern that have contributed to the work described in this lecture include B. Stauffer, A. Neftel, E. Moor. H. Oeschger, J. Schwander, T. Staffelbach and R. Zumbrunn.

REFERENCES

1. C. D. Keeling, R. B. Bacastov and T. P. Whorf, Measurement of the concentration of Carbon Dioxide at Mauna Loa Observatory, Haway, in W. C. Clark (ed.), "Carbon Dioxide Review", Oxford University Press, New York, p. 377:385 (1982)

2. B. E. Lehmann, M. Wahlen, R. Zumbrunn and H. Oeschger, _Applied Physics_, 13, 153 (1977)

3. R. Zumbrunn, A. Neftel and H. Oeschger, _Earth and Plan. Sci. Lett._, 60, 318 (1982)

4. A. Neftel, H. Oeschger, J. Schwander, B. Stauffer and R. Zumbrunn, _Nature_, Vol. 295, N. 5846, 220 (1982)

5. A. Neftel, E. Moor, H. Oeschger and B. Stauffer, _Nature_, 315, 45 (1985)

6. A. Neftel. H. Oeschger, T. Staffelbach and B. Stauffer, _Nature_, Vol. 331, N. 6157, 609 (1988)

7. H. Oeschger and U. Siegenthaler, How has the atmosperic concentration of CO_2 changed", _in:_ "The Changing Atmosphere", F. S. Rowland and I. S. A. Isaksen, eds., p. 5-23, John Wiley and Sons Ltd. (1988)

RESONANCE IONIZATION SPECTROSCOPY

FOR DATING VERY OLD GROUNDWATER

B. E. Lehmann
Physics Institute
University of Bern
Bern, Switzerland

I. INTRODUCTION

When cosmic rays hit the Earth's atmosphere a number of radioisotopes such as ^{14}C, ^{39}Ar, ^{10}Be and ^{39}Cl are produced, which are extensively used in isotope geophysics as natural tracers and for dating various materials from organic substances in archeology to old groundwater and polar ice core samples. The concentration of these trace substances is usually extremely low, requiring very sensitive detection techniques. Low back-ground proportional counters and more recently accelerator mass spectro-metry have successfully been used. With both these techniques, however, it is not possible to detect ^{81}Kr, which is another cosmic-ray produced radioisotope of great potential, especially for use in groundwater- and polar ice core dating. Its half-life of 210,000 years would enable studies of natural archives back to about one million years.

For hydrology the following analytical problem has to be solved: detect 1400 or less atoms of ^{81}Kr per liter of water[1]. The situation is further complicated by isotopic interference with stable krypton isotopes, resulting in a Kr/^{81}Kr-ratio of at least 1.9 x 10^{12}. For practical work it would be best to analyze as small a water sample as possible. Samples of up to 10,000 liters can still routinely be degassed in the field, for instance to extract argon gas for ^{39}Ar dating. With the system described in this lecture which is based on laser resonance ionization spectroscopy it is currently possible to process water samples of approximately 100 liters.

II. RESONANCE IONIZATION SPECTROSCOPY (RIS)

The RIS technique combines the selectivity of optical spectroscopy with the sensitivity of ionization detectors. Tunable lasers excite and ionize atoms of a selected element; ions (or electrons) are subsequently registered with almost unit efficiency. The counting of single atoms in

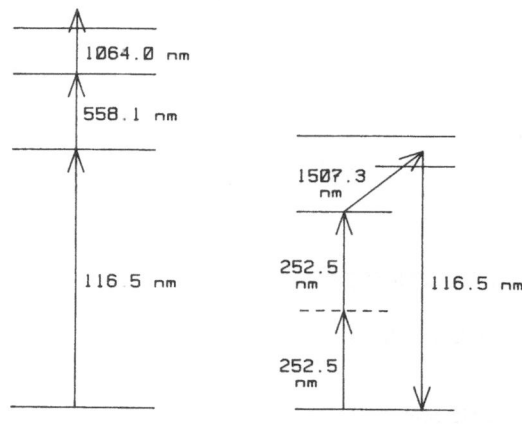

Fig. 1. RIS scheme for krypton (left side)
and four-wave mixing scheme in xenon
for the generation of tunable light
at 116.5 nm (right side).

macroscopic samples has become possible. However, with this ultimate
analytical sensitivity it is immediately clear that any contamination of a
sample can also be measured at extremely low levels. In general it is
therefore important - as was pointed out by other authors in this book -
to combine RIS-techniques with proper sample handling procedures. These
may turn out to be equally as complicated as the actual laser spectroscopy
itself.

For ^{81}Kr-detection both aspects required considerable work in developing a system tailored to the specific problems of this radioisotope:
i) Spectroscopy. It was necessary to find ways of generating tunable
    UV-light at the very short wavelength of 116.5 nm due to the fact that
    krypton as a inert gas has no electronic states below 9.9 eV.
ii) Sample processing. At that short wavelength only pulsed laser systems
    can be operated which have optical bandwidths that are larger than the
    very small isotope effects in the krypton spectrum. Therefore all
    isotopes of krypton are simultaneously ionized by the laser pulses.
    Any isotopic selectivity has to come from combining the RIS ion source
    with a conventional mass spectrometer. For a reasonably small system,
    abundance sensitivities on the order of 1000 are typical. For the
    detection of ^{81}Kr in natural samples the original Kr/^{81}Kr-ratio of
    about 10^{12} therefore has to be lowered in a number of "classical"
    isotope enrichment steps prior to the measurement.

III. RIS OF KRYPTON

Figure 1 shows the two-photon resonant, three-photon ionization
scheme that is used to ionize krypton atoms. To excite the 5s' transition,
tunable light at 116.49 nm is required which is produced in a four-wave

220

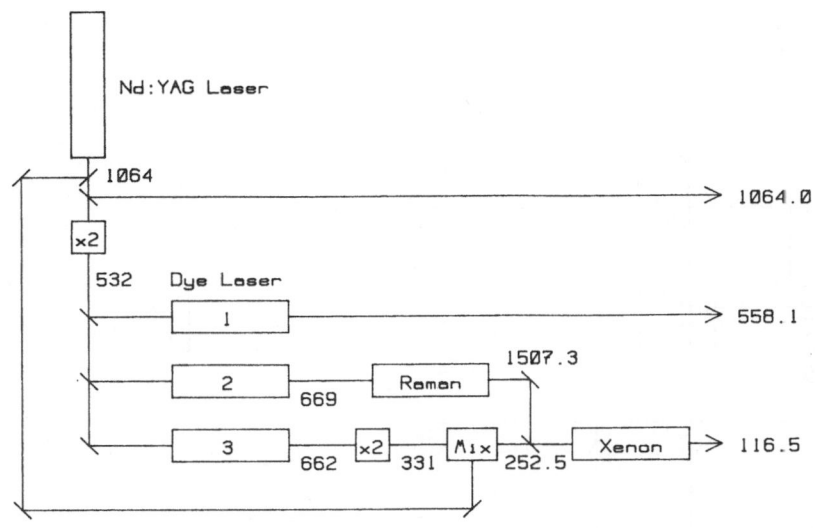

Fig. 2. Schematic view of the laser arrangement for the generation of all wavelengths used in the RIS scheme of Kr.

mixing process in xenon, also outlined in Figure 1. Light from the second
harmonic of a pulsed Nd:YAG laser at 532 nm is used to pump three dye
lasers for generation of the various tunable outputs at 558.1 nm, 669.0 nm
and 662.0 nm as summarized schematically in Figure 2. The tunable near
infrared light at 1507.3 nm is produced as a second Stokes shifted output
in a high pressure hydrogen Raman cell. Tunable VUV-light pulses with
energies up to 0.7 μm are generated with this set-up. More details are
given by Kramer[2] et al., 1983.

IV. RESONANCE IONIZATION SPECTROSCOPY TIME-OF-FLIGHT SYSTEM

The element-selective ion source is combined with a time-of-flight
mass spectrometer as shown in Figure 3. Lasers are used for three
different purposes in this arrangement:

a) Laser annealing. At the end of the sample processing (see below)
krypton atoms are implanted at 10 kV in high purity silicon target. This
chip is then introduced into a vacuum lock where the surface can be hit by
frequency-doubled Nd:YAG radiation. When about 1 J cm^{-2} is absorbed,
silicon melts to a shallow depth of about 1 μm. As soon as the rapid
absorption is over (about 10^{-8} s laser pulses), the crystal begins to
reform. As the solid-liquid interface propagates back to the irradiated
surface, atoms which are more soluble in the liquid compared with the
polycrystalline substrate will be continuously expelled into the liquid
phase and finally leave the surface. The krypton gas sample is then introduced to the main volume for analysis.

b) Laser atom bunching. With the RIS scheme described above all krypton
atoms in an effective detection volume of about 2 x 10^{-4} cm^3 in front of

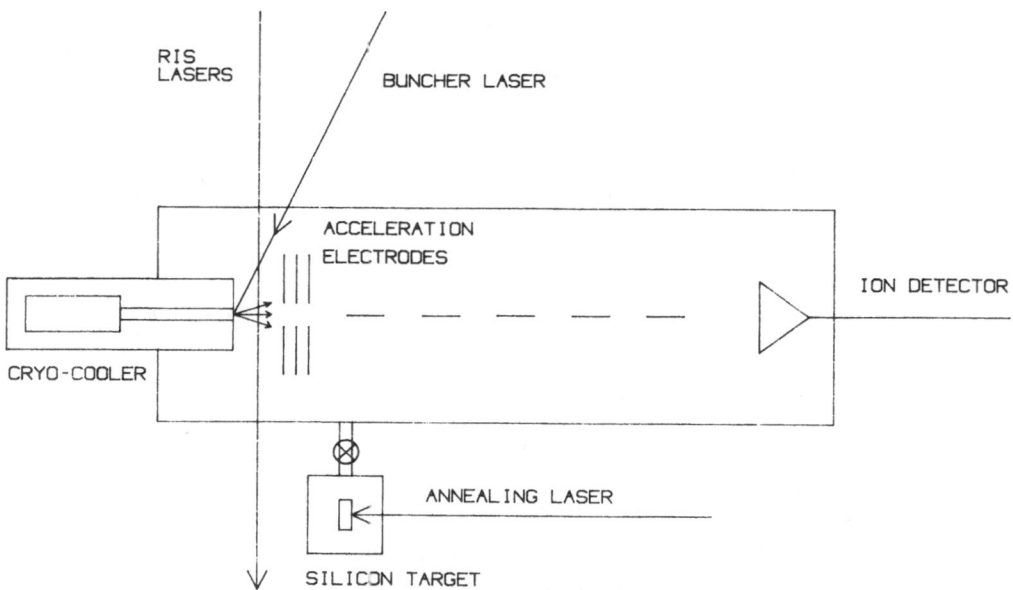

Fig. 3. Resonance ionization spectroscopy time-of-flight system
currently in operation at Atom Sciences Inc. in Oak Ridge.

the ion extractor can be ionized. The time for counting 63% (1-1/e) of the
atoms in a volume of 5000 cm^3 when running all lasers at a repetition rate
of 10 Hz would be an unrealistic long 700 hours. It was therefore
necessary to find a way of concentrating atoms in the RIS-beams[3]. This
arrangement is called the "atom buncher". It consists of a well-defined
cryogenically cooled cold spot of approximately 3 mm diameter which can be
hit by pulses from a flashlamp-pumped dye laser. Krypton atoms are trapped
at the cold spot until they are released by heating with the buncher
laser. This laser also operates at 10 Hz but is fired about 10 μs before
the RIS-lasers to maximize the spatial overlap between the cloud of atoms
escaping from the surface and the pulse of photons crossing from the side.
In practice a reduction of the counting time of a factor of about 1000
results in making krypton atom counting possible in less than one hour.

c) Laser ionization. Laser pulses at 116.5 nm, 558.1 nm and 1064.0 nm that
overlap in space and time in front of the cold spot ionize krypton atoms
as discussed above. With the short laser pulses of 5-10 ns it is particularly important to also verify that the various path lengths on the optical
table are of the same magnitude, since a difference of 30 cm results in a
1 ns time delay.

The sensitivity of the RIS-TOF system is indicated in Figure 4 taken
from Thonnard[4] et al., 1987. The signal at mass 81 in the spectrum represents a total of only 5000 atoms of ^{81}Kr in the closed total volume of the
mass spectrometer of about 6 litres. About 5×10^6 atoms were present in
the sample. Such a detection sensitivity is one to two orders of magnitude

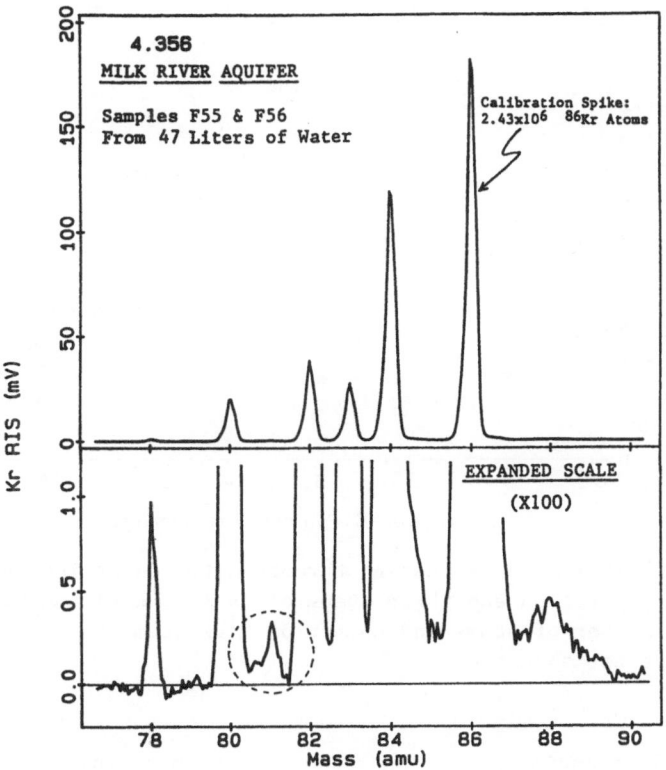

Fig. 4. Example of a [81]Kr spectrum from the RIS-TOF
system of Figure 3. The peak at mass 81
represents approximately 5000 atoms.

below the best conventional mass spectrometers and illustrates the extreme
elemental selectivity that can be realized by RIS techniques.

Additional technical details of the individual components and procedures for counting noble gas atoms with isotopic selectivity are described
by Hurst[5] et al., 1985.

V. SAMPLE PROCESSING

As was mentioned before, a groundwater sample has to go through a
complicated sample processing procedure prior to the analysis in the RIS-
TOF system. Seven steps can be distinguished. The number of atoms after
each can be distinguished. The number of atoms after each of these steps
is indicated in Figure 5. The goal is to sort the few [81]Kr atoms out of
a large water sample with as little loss as possible.

Step 1: water samples of 50-100 liters are necessary for a [81]Kr measurement. Care has to be taken that no contact between the water and the
atmosphere takes place in order to avoid contamination with atmospheric

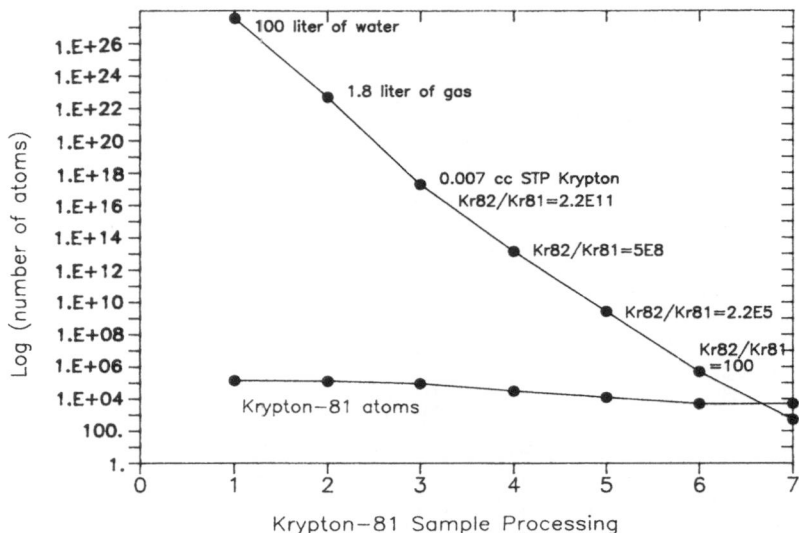

Fig. 5. ^{81}Kr sample processing through water degassing, krypton
separation and three steps of isotope enrichment. Total
number of atoms and number of ^{81}Kr atoms in the sample
after each step.

Kr. The sampling equipment is connected either to an inlet tube down in a
borehole or to an outflow at the surface.

Step 2: degassing of the water is accomplished in the field by spraying
the water into a vacuum system and compressing the evolving gases into
sample cylinders that are returned to the laboratory.

Step 3: krypton separation is achieved by gas chromatography and getter
techniques. The goal is to remove all nitrogen, oxygen, argon, water
vapour and any other gases that are dissolved in the groundwater, without
losing krypton. A sample of approximately 0.007 cm^3 STP of krypton gas is
recovered after this step. It contains 1.9×10^{17} atoms of Kr and 10^5
or less atoms of ^{81}Kr.

Step 4: in a first step of isotope enrichment the sample is processed by a
Wien filter. Ions from a plasma discharge ion source are accelerated to
10 kV. The spatial separation of ions after a flight path of approximately
150 cm is on the order of 1 mm. ^{81}Kr atoms are separated by an 1 mm
aperture plate from the rest of the stable krypton atoms and implanted
into an aluminum coated kapton foil[6]. The enrichment factor for the ^{82}Kr/
^{81}Kr ratio is about 500.

Steps 5 and 6: the sample then has to go through two more steps of isotope
enrichment. A quadrupole mass spectrometer system is used in the present
procedure, and ^{81}Kr atoms are implanted into silicon targets. Enrichment
factors of about 2000 are realized in this step[7].

Step 7: as described above the sample is finally introduced to the RIS-TOF system for ^{81}Kr atom counting.

VI. ACKNOWLEDGEMENTS

The work described in this lecture is based on a long collaboration between groups in Oak Ridge, USA and in Bern, Switzerland. Members of the Photophysics Group at the Oak Ridge National Laboratory that have contributed to the development of a ^{81}Kr technique include G. S. Hurst, M. G. Payne, S. D. Kramer, C. H. Chen, S. L. Allman and R. C. Phillips. At Atom Sciences Inc. in Oak Ridge where the RIS-TOF system was built and where the spectrum of Figure 4 was taken N. Thonnard, R. D. Willis, M. C. Wright and W. A. Davis have worked on this project. At the Physics Institute of the University of Bern the following scientists were involved: H. Oeschger, H. H. Loosli and D. Rauber.

REFERENCES

1. B. E. Lehmann, G. S. Hurst, S. L. Allman, C. H. Chen, S. D. Kramer, M. G. Payne, R. C. Phillips, R. D. Willis, N. Thonnard, H. Oeschger and H. H. Loosli, J. of Geophys. Res., 90, B13, 11547 (1985)
2. S. D. Kramer, C. H. Chen, M. G. Payne, G. S. Hurst and B. E. Lehamn, Appl. Opt., 22, 3271 (1983)
3. G. S. Hurst, R. C. Phillips, J. W. T. Dabbs and B. E. Lehmann, J. Appl. Phys., 55, 5, 1278 (1984)
4. N. Thonnard, R. D. Willis, M. C. Wright, W. A. Davis and B. E. Lehmann, Nucl. Instr. and Methods, B29, 398 (1987)
5. G. S. Hurst,, M. G. Payne, S. D. Kramer, C. H. Chen, R. C. Phillips, S. L. Allman, G. D. Alton, J. W. T. Dabbs, R. D. Willis and B. E. Lehmann, "Report on Progress in Physics", 48, 1333-70 (1985)
6. B. E. Lehmann, D. F. Rauber, N. Thonnard and R. D. Willis, Nucl. Instr. and Methods, B28, 571 (1987)
7. R. D. Willis, N. Thonnard, M. C. Wright and B. E. Lehmann, "Counting ^{81}Kr atoms in groundwater using RIS-TOF", Inst. Phys. Conf. Series 94, 5, 213 (1988)

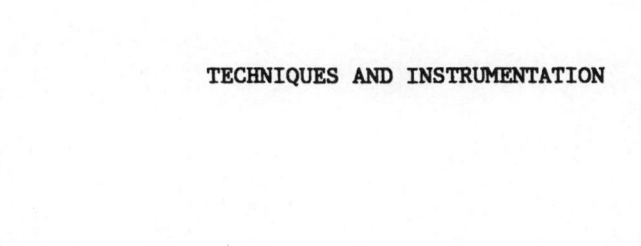

TECHNIQUES AND INSTRUMENTATION

LASER ABLATION FOR MICROANALYSIS

K. Niemax

Inst. für Spektrochemie und Angewandte Spektrosk. (ISAS)
Dortmund, Fed. Rep. of Germany

I. INTRODUCTION

Elemental analysis by direct ablation and atomization of microsamples using pulsed, focused laser beams is a fascinating method which was first applied soon after the first lasers were available[1]. Since that time, many research groups have worked in the field of microanalysis by laser ablation (for reference see, for example, reviews[2,3]). Almost all groups used optical emission spectrometry (OES) of the laser induced plasma to analyse the ablated samples. However, the analytical figures of merit (accuracy, precision and detection limit) were often very poor. The poor analytical results were due to bad reproducibilities of the laser power and to poor laser beam profiles. On the other hand, often the optical detectors were not able to record rapid changing, transient spectra with good time resolution. In this contribution, basic investigations on the ablation process and the optimization of the atomization process will be reported and discussed. The analytical figures of merit will be given for direct optical emission spectrometry of the laser produced plasma and of laser induced fluorescence spectrometry (LIF) using pulsed tunable dye lasers for selective excitation of analytes in the laser microplasma. Furthermore, a way will be shown how the analytical data can be calibrated by internal standardization and the use of standard reference material. For further details see Ref. 4-8.

II. ANALYTICAL ASPECTS OF THE LASER ABLATION AND ATOMIZATION PROCESS

Sample material which has been ablated from a solid by focused laser radiation must be representative for the area hit by the laser beam. If the surface of a solid is heated up slowly by long laser pulses or by a continuous wave laser there is the danger that analytical data are in systematic error due to fractional evaporation. However, if the area under investigation is hit by a very short and powerful laser pulse droplets and particles are ablated in a kind of explosion caused by a plasma ignition.

The stoichiometric composition of the droplets and particles should be representative for the bulk in the area probed. If the sample is ablated into vacuum, as realized in a commercial instrument (LAMA 500/1000), the atomization is very inefficient. Particles and fragments propagate into the vacuum chamber without having the chance of being fully atomized. Moreover, the reproducibility of the number of free atoms (or ions) is not very good. On the other hand, if laser ablation is performed under buffer gas conditions there is also a plasma ignition in the buffer gas above the probed area. This plasma is heated by the laser (inverse bremsstrahlung) to very high temperatures. At the end of the laser pulse the buffer gas plasma starts to recombine. At the same time, the sample material, ablated from the surface and forming a shock wave, starts to penetrate into the very hot buffer gas. In this case, the energy deposited by the laser in the buffer gas is used partly to evaporate the particles and droplets completely. After complete atomization the elemental analysis can be made by OES, LIF or other methods.

Laser ablation of solid samples has to be combined with very sensitive detection methods. For example, if an elemental concentration of 10 μg/g has to be determined in a 10 ng sample, the detection limit of the element has to be 100 fg. However, if the laser is focused tighter and ablates 10 pg only per pulse, the absolute detection limit has to be at least 100 ag. Therefore, the technique of laser ablation has to be combined with the most sensitive analytical method as, for example, LIF and RIMS (resonance ionization mass spectrometry).

III. EXPERIMENTAL ARRANGEMENT FOR OES AND LIF OF LASER PRODUCED SAMPLE PLUMES

The experimental arrangement for OES and LIF of laser ablated sample plumes is shown in Figure 1. The beam of a pulsed laser is focused on the

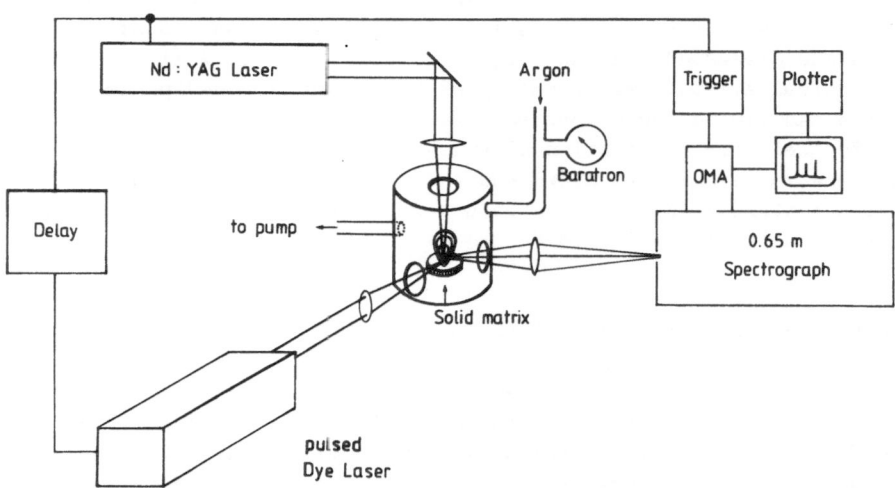

Fig. 1. The experimental arrangement for OES and LIF of laser ablated samples. Figure was taken from Ref. 6.

surface of a solid placed in a pressure-tight chamber under a noble gas
atmosphere. A pulsed dye laser tuned to the resonance line of the analyte
under investigation is fired into the plasma plume with a variable delay.
The fluorescence as well as the emission from the microplasma is measured
by means of a spectrograph and a time-gated optical multichannel analyser.

IV. THE LASER ABLATION PROCESS AND THE ATOMIZATION OF THE MICROSAMPLE

The application of a laser with excellent reproducibilities of the
pulse power and a pure TEM_{00} beam profile is crutial for reliable analytical data. Figure 2 shows LIF data of silicon in dependence on the number of
shots of the ablation laser. The data were taken at different position on
a homogeneous steel sample after "cleaning" the surface by three shots,
respectively. The focus diameter was about 100 μm. One laser pulse removed
a layer of about 1/3 μm thickness from the surface of the solid. The data
are on a straight line. This indicates excellent reproducibility of the
ablation process which is a prediction for good analytical results.

As discussed above, the atomization efficiency is poor when the sample
is ablated into vacuum. On the other hand, the number of free atoms is
also low when the gas pressure is too high. Because of the high viscosity
of the noble gas plasma the chance of redeposition of the ablated material
on the surface of the solid is high. Optimum conditions can be found at
low-pressure conditions. Figure 3 shows a typical time-dependence of the
temperature of a microplasma. The plasma was created by 5 mJ laser pulses
focused on low-alloyed steel samples in a 100 Torr argon atmosphere
(ablation: about 30 ng/pulse). The temperatures were determined from the
emission intensities of iron lines. In the first microseconds after firing
of the ablation laser, the temperature decreases rapidly. Energy deposited

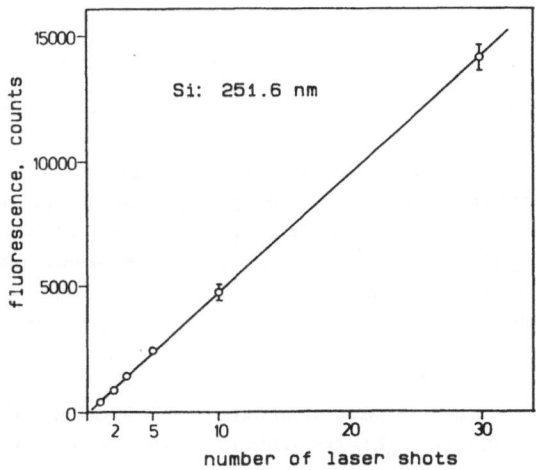

Fig. 2. LIF data of the silicon 251.6 nm line in
dependence on the number of laser shots.

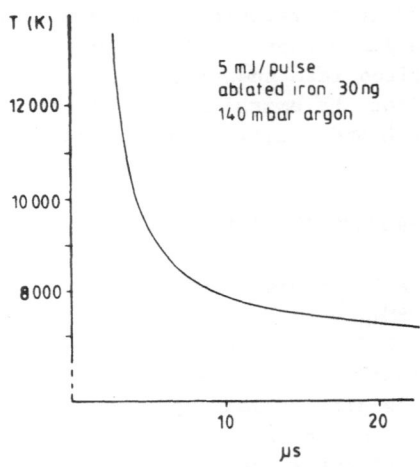

Fig. 3. Time dependence of the plasma temperature.

in the noble gas plasma is consumed by the evaporation of droplets and particles. After about 12 µs the atomization process is completed and the temperature decreases much slower. The final temperature depends strongly on the amount of ablated material and on the matrix[4].

If the atomization process of the ablated material is completed the ratios of elemental concentrations in the vapour should be the same as in the bulk. Therefore, the method of internal standardization should be applicable to OES and LIF. In OES, the ratio of elements can be measured by choosing pairs of emission lines which have about the same excitation emergies. By this procedure the data are independent on the amount of ablated material and on the plasma temperature. In LIF, the ratios of, e.g., ground state atoms are probed directly. In OES as well as in LIF, the degrees of ionization of the atoms which depend on the plasma temperatures have to be low or should be not too different for the matrices under investigation.

V. OES OF MICROPLASMAS

Figure 4 shows emission spectra derived from low-alloyed steel samples with chromium concentrations up to 360 ppm. As can be seen, the Cr emission line was even detectable with the high-purity iron sample (Cr: about 23 ppm). We found typical detection limits in the lower ppm range for most elements in metal matrices. The reproducibility of the measurements good was (2-6%) if homogeneous samples were used. The dynamic range of the measurements was typically four orders of magnitude.

The spectra of Figure 4 were measured with a delay of 5 µs and a gate width of 1 µs. Although the ablated material was still not fully atomized by the microplasma at that time, the good analytical data can be understood having in mind that the matrices (about 98% Fe) were the same. The quantities of ablated sample material as well as the plasma temperatures

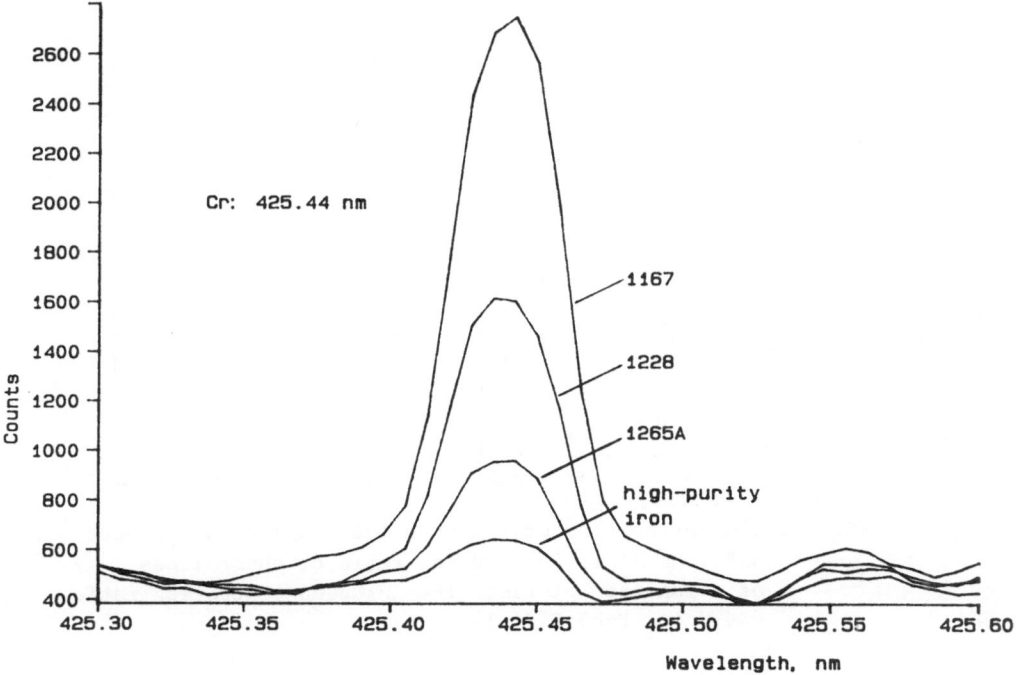

Fig. 4. The Cr 425.4 nm emission line measured with different
NBS standard steel samples. The chromium concentrations
are 360, 160 and 70 µg/g, respectively. (From Ref. 4).

were the same in all measurements. However, strong changes of the
elemental composition in the matrix can change the amount of ablated
material. As a consequence,the temperature of the microplasma will change
and, therefore, influence the emission intensities of the analyte lines.
Differences of analyte concentrations and temperatures in the plasma can
be eliminated by internal standardization using line pairs with similar
excitation energy. However, the evaporation process of sample material in
the plasma has to be completed. Otherwise fractional evaporation may
produce systematic errors. For example, the intensity ratios of Zn/Cu line
pairs vary strongly during the evaporation process of ablated brass
samples (see Figure 5). If the atomization is finished (at about 16 µs, see
Figure 5) the ratios do not change anymore[7].

VI. LIF IN MICROPLASMAS

The detection limits can be improved if the technique of LIF is
applied to the laser microplasma. For examples, detection limits in the
ppb range have been demonstrated for several elements[5,8] probing metal
samples. Absolute detection limits were found to be in the fg range for a
single laser shot in preliminary measurements. The precision (about 1-4%)
as well as the dynamic range of the analyte signals were found to be
excellent when homogeneous samples were analysed.

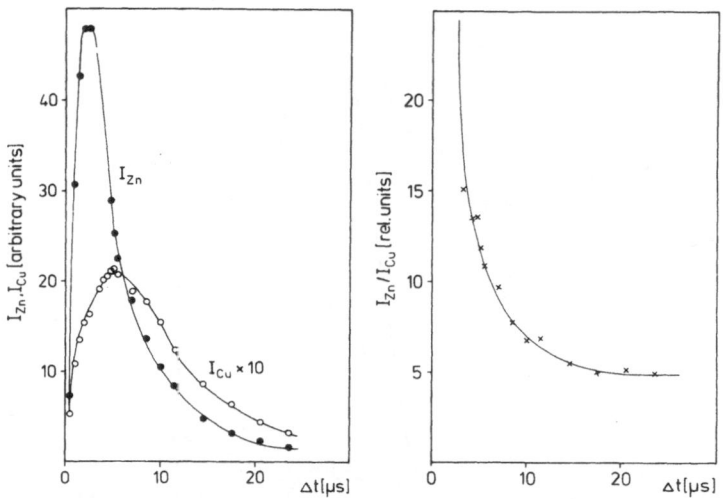

Fig. 5. The intensities of the zinc 468.0 nm (●) and copper
453.1 nm lines (o) and the ratio of these lines
as a function of time. The Zn/Cu concentration ratio
was 0.64 (From Ref. 7).

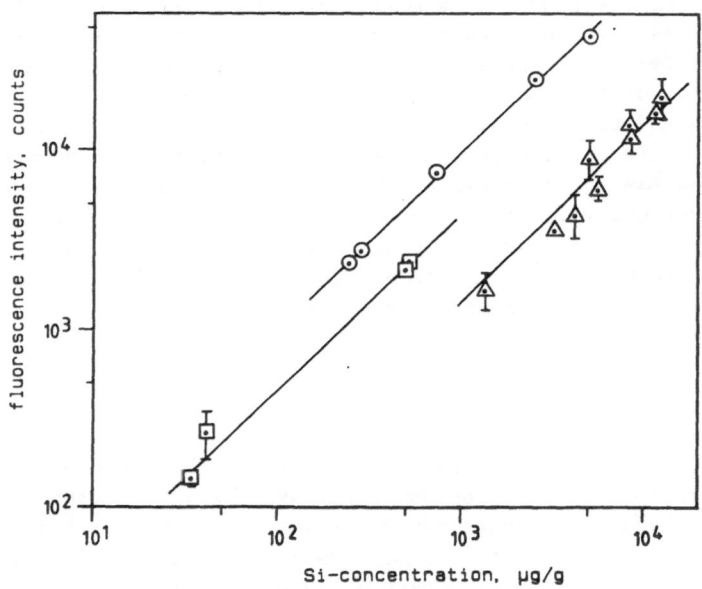

Fig. 6. LIF data of silicon measured in steel (○), copper (□)
and aluminum matrices (△). (From Ref. 8).

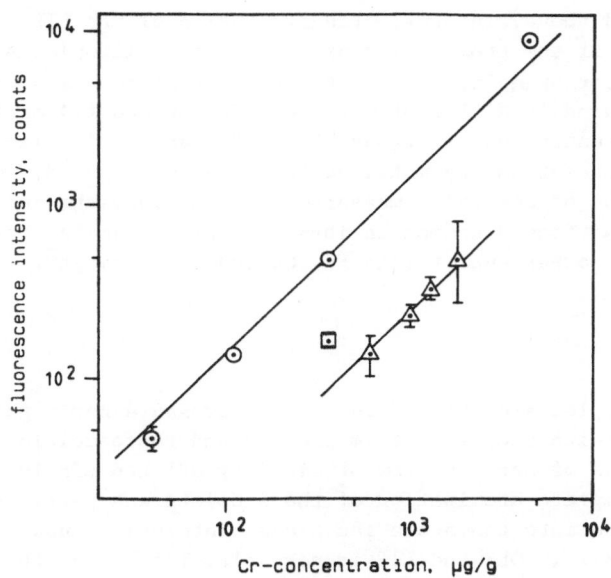

Fig. 7. LIF data of chromium measured in steel (○), copper (□) and aluminum matrices (△). (From Ref. 8).

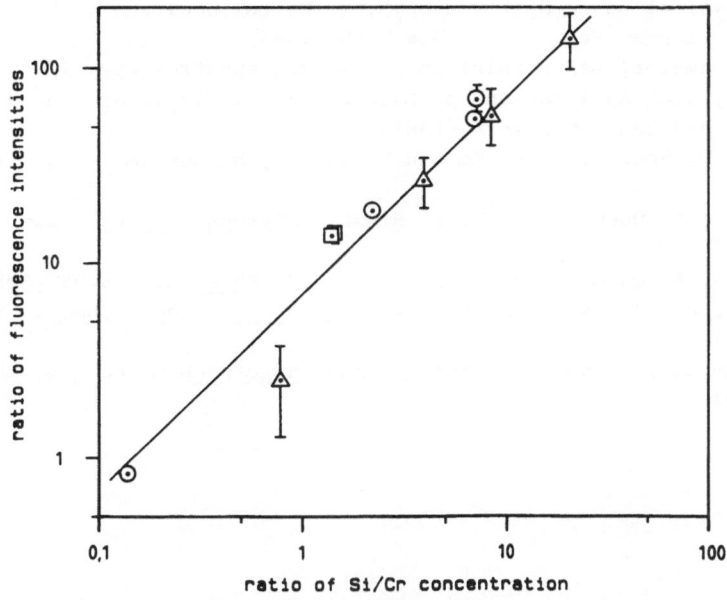

Fig. 8. Ratio of Si/Cr intensities mesured in steel (○), copper (□) and aluminum (△) versus the Si/Cr concentration. (From Ref. 8).

As mentioned above, internal standardization of the LIF signals should also be possible if the atomization process in the microplasma is finished. Figure 6 shows, for example, a plot of ratios of silicon and chromium LIF signals measured in different matrices (Fe, Cu and Al) against the ratio of the concentrations. Although the LIF signals of silicon and chromium are dependent on the matrices (see Figures 7 and 8, respectively), with the exception of one value measured in aluminium, the data of the ratios are not far from a common calibration curve. Similar results have been obtained for other analytes in Fe, Cu and Al matrices[8].

VII. CONCLUSIONS

If solid samples are ablated into a low pressure noble gas atmosphere by pulsed lasers with excellent beam profile and reproducible power, good analytical figures of merit can be obtained by OES and LIF in the microplasma. After complete atomization of the droplets and particles which have been diffused into the noble gas plasma internal standardization of the analyte signals in OES and LIF is possible. This opens the possibility to calibrate analytical signals independent on the matrix by standard reference material. This, in particular, is important in laser microanalysis where the elemental composition may vary dramatically in dependence on the position probed.

REFERENCES

1. F. Brech and L. Cross, Appl. Spectrosc., 16:59 (1962)
2. K. Laqua, Analytical spectroscopy using laser atomizers, in: "Analytical Laser Spectroscopy", S. Martellucci and A. N. Chester, eds., Plenum. Pub. Corp., New York (1985)
3. E. H. Piepmeier, Laser ablation for atomic spectroscopy, in: "Analytical Applications of Lasers", E. H. Piepmeier, ed., John Wiley and Sons, New York (1986)
4. F. Leis, W. Sdorra, J. B. Ko and K. Niemax, Mikrochim. Acta, II:185 (1989)
5. W. Sdorra, A. Quentmeier and K. Niemax, Mikrochim. Acta, II:201 (1989)
6. W. Sdorra, A. Quentmeier and K. Niemax, Z. Phys. D., 13:95 (1989)
7. J. B. Ko, W. Sdorra and K. Niemax, Fresenius Z. Anal. Chem., 335:648 (1989)
8. A. Quentmeier, W. Sdorra and K. Niemax, Spectrochim. Acta B, 45B:537 (1990)

THE FUNDAMENTALS OF TRACE ANALYSIS

USING RESONANT IONIZATION MASS SPECTROSCOPY

K. W. D. Ledingham

University of Glasgow
Department of Physics and Astronomy
Glasgow, Scotland

I. INTRODUCTION

Resonance Ionization Spectroscopy (RIS) is a technology which was
developed during the seventies, principally by Letokhov and his colleagues
in Moscow and Hurst and Payne and co-workers[2] at Oak Ridge in the USA.
Early on it was recognized that RIS was a very sensitive technique which
could detect down to the single atom level[3]. However it took several more
years - until the early eighties - before the methodology was sufficiently
developed that the ultra sensitive analytic technique of Resonance Ionization Mass Spectroscopy evolved (RIMS)[4-7].

Although the samples to be analysed can be in solid, liquid or gaseous
states, the preferred form is solid and both stable and radioactive material
can be assayed. If the sample is a solid then an analysis consists of three
steps (Figure 1):
a) The sample is gasified by either ion sputtering or laser desorption to
   form a plume of principally neutral atoms or molecules.
b) The neutral atoms or molecules in the plume are resonantly ionized
   using one or more laser beams usually pulsed and broadband. For atoms,
   this step is elementally very selective with isotopic selectivity being
   carried out in the next step (c). Isobaric interferences which plague
   conventional mass spectroscopy are largely eliminated at this stage.
c) Finally the ions created are mass analysed, normally using a time-of-
   flight mass spectrometer which complements ideally the pulsed aspect of
   the ion source.

Each of these steps will be described in the next section (procedure),
followed by typical results and limits of sensitivity and finally in the
conclusions possible improvements will be discussed.

Optoelectronics for Environmental Science, Edited by S. Martellucci and
A.N. Chester, Plenum Press, New York, 1990

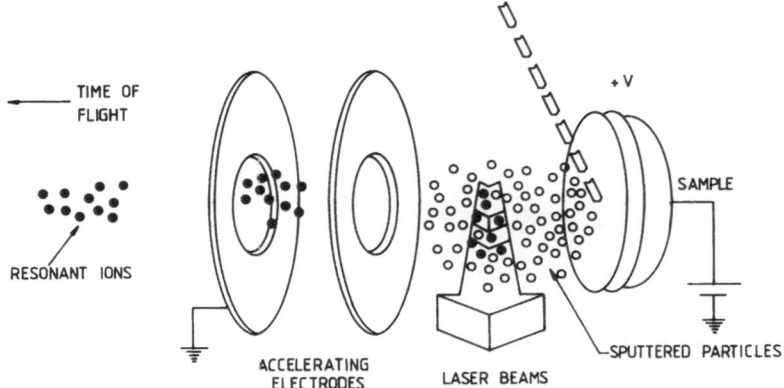

Fig. 1. The three steps involved in RIMS: (a) sputtering/
ablation; (b) resonant laser ionization; and,
(c) extraction and mass identification.

II. PROCEDURE

II.1. Ion Sputtering and Laser Desorption

Either of these processes for creating neutral atoms or molecules has
pros and cons. Ion sputtering is preferred if profiling is involved, i.e.
determining the concentration of a particular element as a function of
depth below the surface and is invariably used for analysing semiconductor
materials. On the other hand laser desorption is a soft ablation procedure
for liberating molecules from a surface intact, i.e. with minimal fragmen-
tation which often occurs with ion sputtering. Hence laser desorption is
popular in the analysis of organic molecules especially when preserving
the parent mass peak is important.

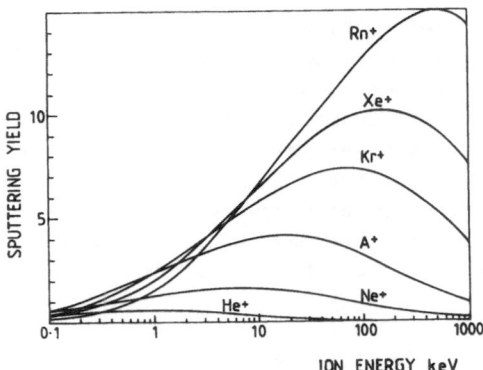

Fig. 2. Calculated sputtering yields for aluminium
as a function of ion energy and ion mass[9].
(Reproduced by permission of the authors
and Academic Press).

II.2. Ion Sputtering

The most important aspect of ion sputtering is to maximise the sputtering yield, i.e. the number of neutral atoms created from the sample per incident ion. The yield is a complicated function of many parameters among which are the incident ion's mass and energy as well as the incident angle, i.e. the angle between the beam and the normal to the sample[8]. A comprehensive treatment of ion sputtering is given in ref. 8 dealing with Secondary Ion Mass Spectrometry (SIMS), a technique which has been established for many years. SIMS employs a similar technology to RIMS but analyses the sputtered ions created by the incident beam rather than the neutrals - a point to which we shall return. Figures 2 and 3 show how the sputtering yield varies as a function of the incident ion's mass, energy[9] and angle of incidence[10]. Clearly an argon ion gun with 5-10 keV energy fitted at an angle of incidence of about 60-70° to the normal is a sensible instrument choice for sputtering and maximizes the sputtering yield at about 5. Typically pulsed ion guns have currents of about a few μamps in pulses of about 1 μsec duration, and hence the number of sputtered neutrals per pulse per μamp is ~ 3 x 10^7.

II.3. Laser Desorption (Ablation)

When a laser beam strikes the surface of the sample, the temperature is immediately elevated and a plasma of ions and neutrals is formed.

The mass ablation coefficient (m) in kg $cm^{-2}s^{-1}$ has been determined experimentally by Fabbro et al.[11] and Goldsack et al.[12] and allows one to

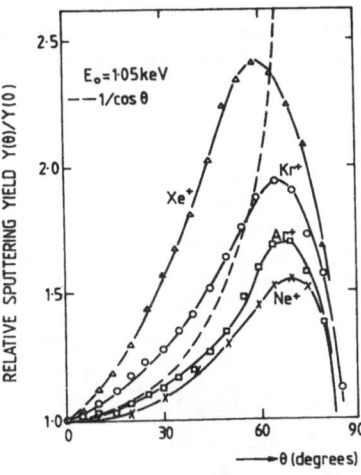

Fig. 3. Relative sputtering yields (θ°/0°). For Ne⁺, Ar⁺, Kr⁺ and Xe⁺ as a function of angle of incidence (θ) to the normal[10]. (Reproduced by permission of the authors and Springer-Verlag).

calculate the number of atoms ablated from the sample surface (assuming the atoms are created free[6]):

$$m \sim 110 \left[\frac{\phi_a}{10^{14}} \right]^{1/3} \frac{1}{\lambda^{4/3}} \tag{1}$$

where λ is the laser wavelength in μm and ϕ_a is the laser energy flux in W cm^{-2}. A Nd:YAG laser is normally used as the ablation laser operated at the 1.06 μm fundamental wavelength, doubled (530 nm), tripled (355 nm) or quadrupled (266 nm). Operating at the fundamental or doubled wavelengths results in the most stable conditions and with a laser fluence of ~ mJ/mm^2, a plasma is created without space change problems. Using equation (1) with a Nd:YAG operated at 1 μm with a pulse length of typically 10 ns, a fluence of 1 mJ/mm^2, the number of atoms of mass 50 amu ablated per pulse is about 6 x 10^{14}.

It is clear that a typical laser pulse desorbs many more atoms than a typical ion gun pulse.

Finally in this section two disadvantages of the SIMS technique should be mentioned which are not shared by RIMS and should make the latter technique potentially a more quantitative ultra trace analytical technique. Firstly SIMS analyses the ion fraction of the ejected particles which is often 10^{-2} or less and secondly the ion fraction is a sensitive function of the sample matrix. RIMS on the other hand examines the neutral flux which should dramatically reduce matrix effects and improve the prospects for quantitative analysis.

II.4. Resonant Ionization of Neutrals

Resonant ionization spectroscopy is a photoionization method by which the neutral atoms in the plume above the sample are ionized by the absorp-

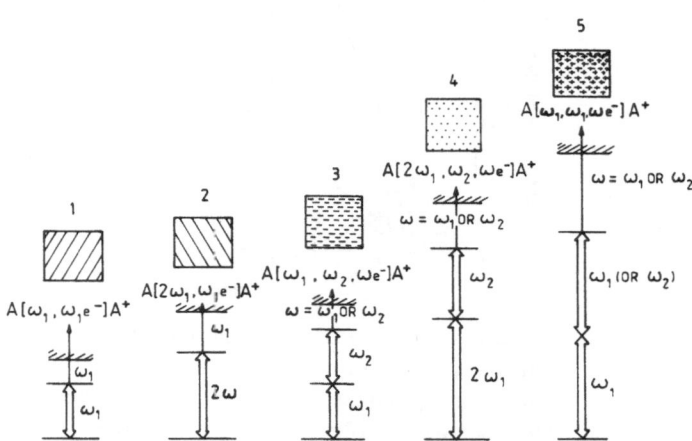

Fig. 4. The five ionization schemes proposed by Hurst et al.[5].

tion of two or more photons from tunable dye lasers. One or more of the adsorbed photons energetically match transitions between quantum states of an atom or molecule and hence the process is resonant. Hurst et al.[2] have proposed five basic ionization schemes according to the relative energy position of the intermediate states to the continuum. These are shown in Figure 4. In the first scheme a level exists at an energy more than $\frac{1}{2}$ I (Ionization Potential) and hence the atom can be ionized by a resonant two photon process (same colour). In scheme two the output of the dye laser (λ) must be frequency doubled ($\lambda/2$) to excite the atom and then ionized by a photon (λ) from the original beam. The other three schemes involve the absorption of three photons. Figure 5 shows a diagram of the periodic table with one of the five schemes being ascribed to each element. It must be emphasized that these are only suggested schemes for ionizing a particular element and a study of the atomic energy level tables[14] will suggest many more. It can be seen from Figure 5 that many elements can be ionized using a single dye laser which has a frequency doubling capacity including most of those important to the semiconductor industry. For complete elemental coverage, one requires a large pump laser (Excimer or Nd:YAG) and two dye lasers, one of which is frequency doubled.

To maximise the number of ions formed, the ionization process should be saturated, i.e. each neutral atom in the laser beam will be ionized. Figure 6 shows typical atomic excitation and ionization cross sections. In Figure 6(a) the excitation step can be saturated with laser fluences of about 1 $\mu J/cm^2$ [15] while the ionization step requires fluences of about 100 mJ/cm^2. To reach such fluences the output of most commercially available dye

I	II	III	IV	V	VI	VII	VIII			0
1H										
3Li	4Be	5B	6C	7N	8O					
11Na	12Mg	13Al		15P	16S	17Cl				
19K	20Ca	21Sc	22Ti	23V	24Cr	25Mn	26Fe	27Co	28Ni	
29Cu	30Zn	31Ga	32Ge	33As	34Se	35Br				36Kr
37Rb	38Sr	39Y	40Zr	41Nb	42Mo	43Tc	44Ru	45Rh	46Pd	
47Ag	48Cd	49In	50Sn	51Sb	52Te	53I				54Xe
55Cs	56Ba		72Hf	73Ta	74W	75Re	76Os	77Ir	78Pt	
79Au	80Hg	81Tl	82Pb	83Bi	84Po					
87Fr	88Ra									

57La	58Ce	59Pr	60Nd	61Pm	62Sm	63Eu	64Gd	65Tb	66Dy	67Ho	68Er	69Tm	70Yb	71Lu
			92U			95Am			90Es					

Fig. 5. The periodic table with each element being ascribed an ionization scheme described in Fig. 4.

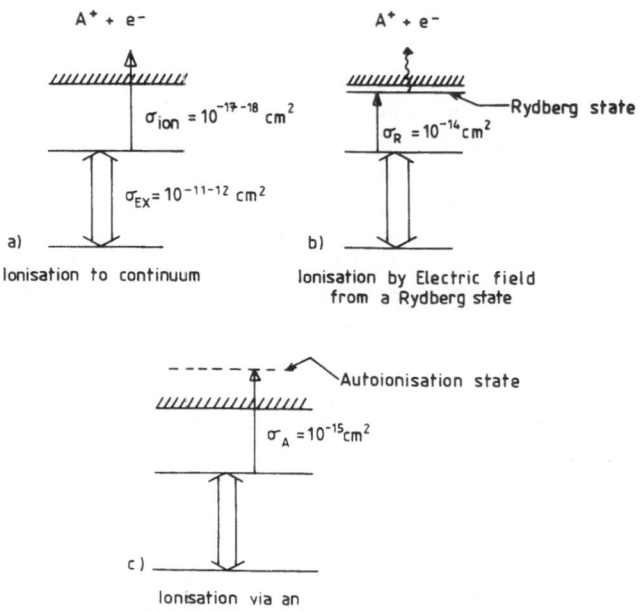

Fig. 6. Different ionization processes: (a) a two photon ionization
process; (b) the atom is excited a Rydberg state and finally
ionized by a pulsed electric field; (c) the final step via
an autoionization level.

lasers must be moderately focused which has the disadvantage of reducing
the interaction volume and hence the number of atoms which can be
analysed. Two other ionization procedures have been suggested by Bekov and
Letokhov[4] which alleviate this problem and are shown in Figure 6(b) and (c).
In Figure 6b the atom is excited to a Rydberg level close to the continuum
and can be ionized with 100% efficiency by the application of a pulsed field
of about 10^4 V/cm. The other method is to ionize the atom via autoionization states (Figure 6c). The rate limiting step in (b) and (c) has a cross
section some two orders of magnitude larger than in (a) and hence requires
much lower fluences to reach saturation.

Finally to maximise the overlap between the sputtered/ablated atoms
and the ionizing lasers, the RIS beams should be unfocused and pass as
close to the sample as possible. The timing between the sputtering/abla-
tion and ionizing pulses is also important with the latter being triggered between 0-2 μs after the completion of the former[13].

II.5. Ion Detection

After ionization, the ions are extracted and passed into a time-of-
flight (TOF) mass spectrometer. Although quadrupole and magnetic sector
mass spectrometers have been used in RIMS, TOFs are usually employed since
all masses can be analysed in one pulse and also the transmission through

242

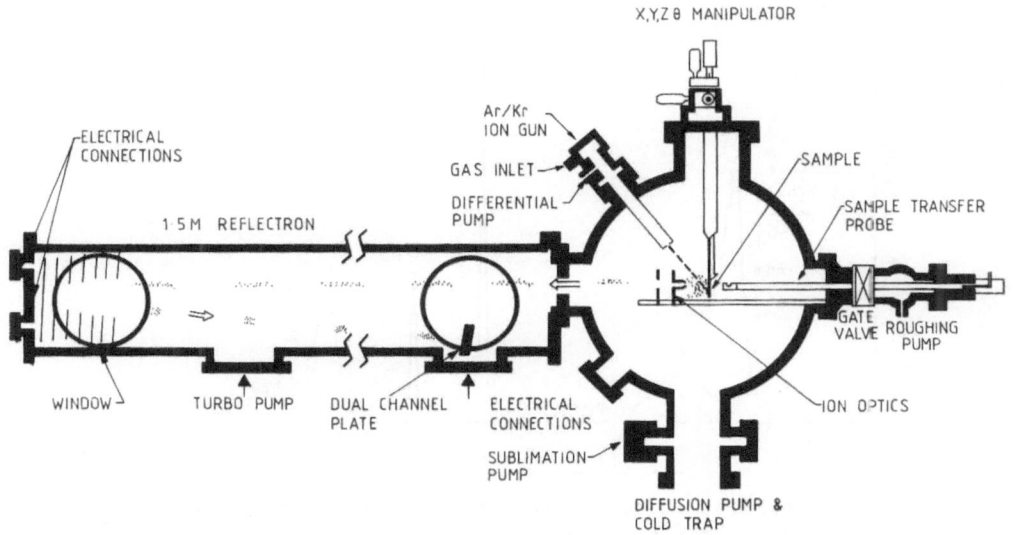

Fig. 7. The Glasgow Resonant Ionization Mass Spectrometer.

the spectrometer is generally superior to other types. The ion extract
optics and a time-of-flight mass spectrometer of reflectron type is shown
in Figure 7. The principal factor which limits the resolution of a conventional TOF spectrometer is the spread of initial ion energies in the
sputtering/ablation process. This spread of ion energies can be compen-
sated using a reflectron TOF[16] in which higher energy ions penetrate deeper
into an electrostatic ion reflector and hence experience a longer flight
time than ions of a lower energy. Thus the spread in initial ion energies
which result in a spread of the arrival times of the ions at the detector
is considerably reduced. The FWHM resolution of the above system is about
1000 for ions of about 40 amu. Finally the ions are detected by a Galileo
double microchannel plate detector with a 0.2 ns rise time. The flight
time for ions of about 40 amu is typically about 50 μs for the overall
drift length of 3 m operated with extract voltages at about 2 kV. A thin
wire (0.005 cm in diameter) follows the ion flight path through the flight
tube providing an electrostatic guide for the ions increasing the transmission of the mass spectrometer. This wire is normally operated at about
-10 V. The sample chamber is pumped by an oil diffusion pump fitted with a
cold trap and a titanium sublimation pump having a total pumping speed of
over 800 1/sec which can maintain a base pressure of less than 10^{-9} Torr.

The data acquisition system measures and stores mass spectra and
laser probe energies on a pulse to pulse basis. A LeCroy 2261 transient
recorder coupled to a IBM PC AT forms the centre of the system. Ion
signals from the detector are digitised by the transient recorder which
provides 640 time channels (11 bit resolution) each of 10 μs width. This
provides time spectra of width 6.4 μs which can be delayed at will by a
Stanford precision pulse generator to give complete mass coverage.

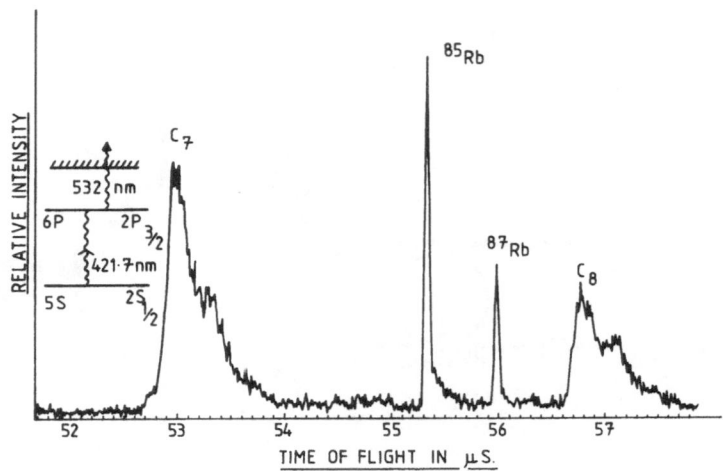

Fig. 8. Rb signal at 15 ppm concentration in NBS coal sample
SRM 1632A. The broad peaks C_7 and C_8 are carbon clusters
produced in the ablation process and the background
corresponds to a Rb concentration of < 1 ppm.

III. RESULTS

The sensitivity of the instrument can be tested using NBS standard
materials. NBS standard coals and ashes (SRM 1632A, 1633, 1645) contain
many elements at ppm concentrations[17] as well as having a difficult matrix
problem with the ablation of a range of carbon clusters and hence is a
stringent test of the capability of the technique. The sample stub was
2.5 cm in diameter with an inset 1 cm in diameter and 2 mm deep in the
middle of the stub. A well homogenised pellet of 50% SRM 1632A and 50%
high purity pelletable graphite to make the sample conducting was placed
in the inset. Conducting samples reduce space charge problems which limit
resolution. Figure 8 shows a Rb(IP 4.18 e V) signal at 15 ppm concentration. An unfocused laser beam of wavelength 532 nm and fluence 1 mJ/mm^2
was used for ablation purposes. The neutral Rb atoms were ionized using a
two step process: a photon from stilbene 420 dye to excite (420.7 nm,
120 μJ focused) and a doubled photon Nd:YAG laser (532 nm, 7 mJ unfocused)
to augment the ionization step. The two laser beams passed through the interaction volume collinearly and were triggered to ionize the neutral plume
about 1 μs after the ablation pulse. The broad peaks to left and right of
the Rb peaks are C_7 and C_8 clusters produced by unsuppressed ablations
ions. The background level in Figure 8 corresponds to a minimum detection
level of Rb in SRM 1632A coal of < 1 ppm.

In the section on procedure it was stated that RIMS was likely to
have fewer matrix problems than SIMS. Figure 9 shows the results of a SIRIS
analysis (sputter induced RIS) to detect silicon in different matrices[18].
The graph is linear with concentration indicating that at least with these
samples the matrix problems are small.

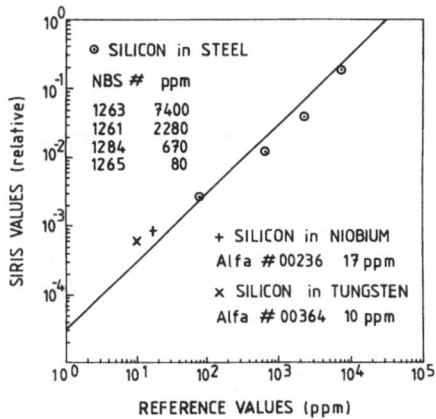

Fig. 9. Comparison of SIRIS measurement using certified [18] reference values of silicon in different matrices. (Reproduced by permission of the authors and the Institute of Physics).

In this paper only analyses at ppm trace concentrations have been discussed and this is due to the fact that there are few standards available at sub ppm levels. The potential however for RIMS is enormous with ppb sensitivity and even parts in 10^{12} being claimed for sensitivity for specific elements[19]. It is going to take some years however before analyses at these levels are routine.

IV. CONCLUSIONS

RIMS has been shown to be a trace analysis technique of considerable potential both qualitatively and quantitatively. Already several commercial firms are advertising RIMS instruments of various types and in addition established laser microprobes can now be fitted with a post ablation ionization capacity both resonant or non resonant. One of the difficulties which instrument manufacturers identify in the RIMS technology is the use of dye lasers with their inherent bulkiness and operating difficulties. With the development of solid state tunable lasers (Ti sapphire or optical parametric oscillators) to replace the dye lasers it is likely that RIMS will soon reach its full potential and be accorded similar acceptance as SIMS.

V. ACKNOWLEDGEMENTS

The author would like to acknowledge the considerable contribution of his colleagues and students in this research.

REFERENCES

1. "Laser Analytical Spectrochemistry", V. S. Letokhov, ed., A. Hilger, Bristol, England (1986)
2. G. S. Hurst and M. G. Payne, "Principals and Applications of Resonances Ionization Spectroscopy", A. Hilger, Bristol, England (1988)
3. G. S. Hurst, M. Y. Nayfeh and J. P. Young, _Phys. Rev._ A15:2283 (1977)
4. G. I. Bekov and V. S. Lekothov, _Appl. Phys._ B30:161 (1983)
5. G. S. Hurst and M. G. Payne, _Spectrochim. Acta_ 43B:1407 (1988)
6. J. D. Fassett and J. C. Travis, _Spectrochim. Acta_ 43B:1407 (1988)
7. D. M. Lubman, _Anal. Chem._ 59 31A (1987)
8. A. Benninghoven, F. G. Rudenauer and H. W. Werner, "Secondary Ion Mass Spectrometry - Basic Concepts, Instrumental Aspects, Applications and Trends", John Wiley & Sons, New York, Chichester, Brisbane, Toronto, Singapore (1987)
9. P. D. Townsend, J. C. Kelly, N. E. W. Hartley, "Ion Implantation, Sputtering and their Applications", Academic Press London, p. 122 (1976)
10. H. Oechsner, _Z. Phys._ 261:37 (1973)
11. R. Fabbro, E. Fabre, F. Amiranoff, C. Carban-Labaune, J. Virmont, M. Weinfield and C. E. Max, _Phys. Rev._ A26:2289 (1982)
12. T. J. Goldsack, J. D. Kilkenny, B. J. MacGowan, S. A. Veats, P. F. Cunningham, C. L. S. Lewis, M. H. Key, P. T. Rumsby and W. T. Toner, _Opt. Comm._ 42:55 (1982)
13. F. M. Kimock, J. P. Baxter, D. L. Pappas, P. H. Kobrin and N. Winograd, _Anal. Chem._ 56:2782 (1984)
14. C. E. Moore, "Atomic Energy Levels", NBS Circular 467, US Government Printing Office, Washington DC (1948)
15. R. P. Singhal, A. P. Land, K. W. D. Ledingham and M. Towrie, _J. Anal. Atomic. Spectroscopy_ 5:599 (1989)
16. B. Mamyrin, V. I. Karataev, D. V. Shmikk and V. A. Zagulin, _Sov. Phys._, JETP 37:45 (1973)
17. E. S. Gladney, _Chimica Acta_ 118:385 (1980)
18. J. E. Parks, D. W. Beekman, L. J. Moore, H. W. Schmitt, M. T. Spaar, E. H. Taylor, J. M. R. Hutchinson and W. M. Fairbank Jr., "Progress in Analysis by Sputter Initiated Resonance Ionization Spectroscopy", RIS 86 Swansea IoP Conf. Ser. 84, 157 (1986)
19. G. I. Bekov, V. S. Letokhov and N. V. Radaev, _J. Opt. Soc. Am._ B/Vol.2, 1554 (1985)

PRODUCTION AND DETECTION OF SELECTIVELY LASER IONIZED SPECIES

F. Giammanco

Dipartimento di Fisica
Università di Pisa
Pisa, Italy

I. INTRODUCTION

Resonant multiphoton ionization is a general method either for analyzing the chemical composition of samples (trace analysis) or for inducing modification of isotopic natural abundance (isotope enrichment). Selectivity is achieved by a multiple-step process in which different laser beams are tuned to resonant transitions of the element of interest. Since the steps involved are, in general electric-dipole transitions, pulsed lasers are required to saturate resonant transitions and, thus, to achieve a high degree of ionization. The ions produced are accelerated by an external electric field, and then fed into a time-of-flight (TOF) mass spectrometer. This simple scheme can be applied to a wide range of molecular or atomic samples[1], providing that the resonant transitions belong to a range of the electromagnetic spectrum in which laser sources are available. In the case of samples whose first resonant step is in the VUV region, e.g. noble gases, excitation by electron impact into metastable states may be combined with selective laser ionization[2].

Applications of this method to the detection of small numbers of specific isotopes benefits a wide range of pure and applied sciences, including weak-interaction physics, solar neutrino research, geological dating, groundwater dating and so on[3].

The analysis of traces is often complicated by the presence of a large number of background ions due to non-selective off-resonance multiphoton ionization. In fact, in many cases the ratio between the probability of the off-resonance process and of the resonant selective ionization is not sufficient to reduce the concentration of the ionized background material to values comparable with the concentration of the element under study. Unselected ions can be distributed over the mass spectrum as an almost uniform background due to collective electron ion interactions, and this reduces sensitivity of the detection. Under

Optoelectronics for Environmental Science, Edited by S. Martellucci and
A.N. Chester, Plenum Press, New York, 1990

specific conditions depending on the type of elements to be identified, a technique based on excitation to high Rydberg levels, followed by delayed ionization by a strong electric field, has been applied successfully[4] to improve the sensitivity up to $\cong 10^{-10}\%$. Unfortunately, this technique cannot be extended to the case in which a particular element must be separated from a background of atoms whose atomic weights and structures of levels are close, e.g. detection of ^{81}Kr in groundwater and rock dating[3]. Moreover, in the case of noble gases, the unselective electron impact step produces a significant number of ions of the most abundant isotopes, since the ionization potential is close to the first excited state energy.

Preparation of samples for selective analysis may constitute an additional source of background ions. In fact, in the case of continuous vaporization, a lot of material is not processed due to the finite laser repetition rate. Thus, pulsed methods of vaporization synchronous with the ionizing laser beams have been proposed[5], in which an intense pulsed laser or electron beam induces phase transition of solid samples. In this case, a large amount of ionized species screens the external field, thus reducing the collection efficiency.

Problems induced by a high value of charge density are well known in isotope enrichment experiments, especially in their application to produce fuel for nuclear plants, where the cost of the process depends strongly on the concentration of the isotope of interest extracted per laser shot[1]. High selectivity is required to suppress the background of ions of the most abundant species (e.g. the natural ^{235}U concentration is $\cong 10^{-3}$ times the ^{238}U one), and the extraction time must be shorter than the charge exchange time constant. For a typical isotope separation experiment in natural Uranium, a ^{235}U charge density of 10^{11} ions/cm^3 is required to assure an advantage for the method, whose corresponding density of neutral ^{238}U atoms is $\cong 10^{14}$ atoms/cm^3 (see Ref. 1). Under these conditions, ^{235}U ions must be collected in a time shorter than $\cong 10$ μs to prevent the charge exchange between ^{235}U ions and ^{238}U atoms from destroying the selectivity of the ionization process[1]. Fast ion extraction is limited by the collective self-generated plasma field due to the different electron-ion mobility that screens the external field. Although this phenomenon is known in the charge density range of the Uranium experiments, only recently several authors[6-10] who dealt with experiments devoted to laser-matter interactions, in a lower charge density range (10^7-10^{10} ions/cm^3), noticed the importance of such "space-charge" effects in modifying the properties of ionized yields, as determined by ionizing sources.

Collective effects depending on the electron-ion mutual interaction in laser multiphoton ionization (MPI), have recently been highlighted and analyzed[11-13]. In particular, a great deal of attention has been directed to their perturbative influence during the collection phase of the initially produced charges. The time-space behavior of charges could be strongly affected by the appearance of a macroscopic electric field, due to the different mobilities of electrons and ions. Hence, the noticeable modifications of the parameters (numbers density, kinetic energy of electrons, ion time-of-flight distribution and so on) of the initially produced charges induce corresponding deviations of the observed dependences on the experimental parameters (laser power, external field, neutral density) from those expected as a consequence of the selected laser-matter interaction. The experiment of Ref. 12, concerning three-photon ionization of a Na beam in the range of experimental parameters common to most MPI experiments, was specifically carried out in order to analyze the main features of the charge evolution. The influence of the self-generated electric field was investigated through an analysis of the electron current and ion yield dependences on the externally applied electric field, laser power and focussing conditions, at different initial gas densities.

The theoretical model, based on a three-dimensinal fluid description of the coupled motion[13] of electrons and ions and also including the production phase during the laser-gas interaction time, represents an extension of the analytic method of solution presented in Ref. 11, and was able to give a quantitative description of the observed dependences.

Collective effects were also observed under completely different experimental conditions, namely during the pulsed injection of a non relativistic (on the order of keV energies) eletron beam in a low density $(10^{-5}$ Torr) gas[14,15]. In this case, a transient stage precedes the steady state, during which the produced charges are completely neutralized. During this transient stage there is a progressive increase of the ion density in the gas, as ions are produced in electron-atom collisions. During the time interval characterizing this stage there is a monotonic transition from a completely unneutralized beam, which is therefore highly divergent in absence of external field, to a steady state beam. What is experimentally observed is a clearly expressed dynamic ion focusing of the electron beam. This focusing was described[14] as an increase in the axial beam current density to the point at which, some time after the beginning of beam injection into the gas, this current density was far greater than the steady-state current density of the beam.

In this chapter, the observations of relevant collective effects in the process of electron impact (EI) ionization of low density atomic and molecular gases (pressure 10^{-6}-10^{-4} Torr) is reported. In particular, there has been observed a focusing ion effect on the electronic beam very similar to that reported in Ref. 14, but at electron energies ($\cong 40$ eV) about 100 times smaller than those characterizing the beams of Refs. 14 and 15 ($\cong 1$ keV), and with an electron beam-gas interaction time ($\cong 10$ nsec) that does not allow the establishment of a steady-state condition. This focusing effect results, in the present case, in an increase of the axial current measured by an electron detector whose radius is smaller than the radial width of the unperturbed eletron beam, which monitors the electron beam current after it has passed through the target gas.

The observed effect shows a strong dependence on the gas pressure, i.e. on the degree of ionization produced. In particular, the peak and the total charge of the electron current exhibit a sudden growth over a very small range of pressure variation and subsequently the behavior is quite flat, except for a non-regular modulation.

The total ion count rate exhibits a change in slope over the same range of pressure. Moreover, the shape of the peaks in the TOF spectrum of the produced ions changes as a function of the gas pressure, showing a growing tail in the region of increasing time of flight.

The results of the EI experiment are compared with those obtained in the MPI experiment of Ref. 12, with special regard to the influence of collective effects on ion detection. Finally, the theoretical model of Ref. 11 and 13 is applied to describe the main features of the EI ionization experiment, and to carry out a simple estimation of the charge threshold at which collective effects become important.

II. ELECTRON IMPACT IONIZATION EXPERIMENT

The experimental set-up is similar to that used in previous experiments[16-18] and thus only a short description is given here.

The TOF ion spectrometer uses a 40 cm long drift tube placed perpendicularly to the electron beam axis. The ions are accelerated at the entrance of the TOF tube by three gaps, across which three different electric fields are applied. The electron beam is produced by a pulsed electron gun that includes an indirectly heated oxide cathode and two electrostatic lenses. The mean kinetic energy of the electrons is (37.5 ± 0.2) eV. The value of the current peak is 100 nA, with a pulse duration of 10 ns at a repetition frequency of 4 kHz. The radial width of the beam in the focal region, which corresponds to the interaction region, is 0.1 cm. The electron beam is monitored by an electron multiplier whose first dynode is positioned 2 cm from the interaction region and whose diameter is 0.25 cm. The electron current signal is fed into a fast (<1 ns rise time) oscilloscope (Tektronix 7844) equipped with a digitizing camera (Tektronix DCS 01) which is interfaced with a personal computer for data storage and analysis.

The gas target, a mixture of Ar and CO_2, is introduced in the ion source through a 10 μm diameter stainless needle injector fed by a variable, programmable, self-controlling leak. The whole apparatus is contained in a stainless steel chamber, with a residual gas pressure (after baking at 200-300 °C) of about 10^{-8} Torr. The overall working pressure in the vacuum vessel during the measurements ranges between 10^{-6} and 10^{-4} Torr, while the local pressure, just above the needle injector, is about 100 times higher. This factor was estimated in previous MPI experiments by measuring with visible laser light the ionization rate both with and without the needle injector, in the same experimental apparatus. The value obtained in the case of laser ionization is corrected for the different interaction volume characterizing the e-beam ionization experiment. The final estimation factor of 100 can be taken with a confidence of 20%. Thus, although on the horizontal axis of Figures 1 and 2 the values correspond to the overall gas pressure in the vacuum chamber, both for cases (a) and (b), the value of 100 ($\pm 20\%$) has to be used as a scale factor for obtaining the actual local pressure characterizing the points of the curves labeled with (a) in the above figures.

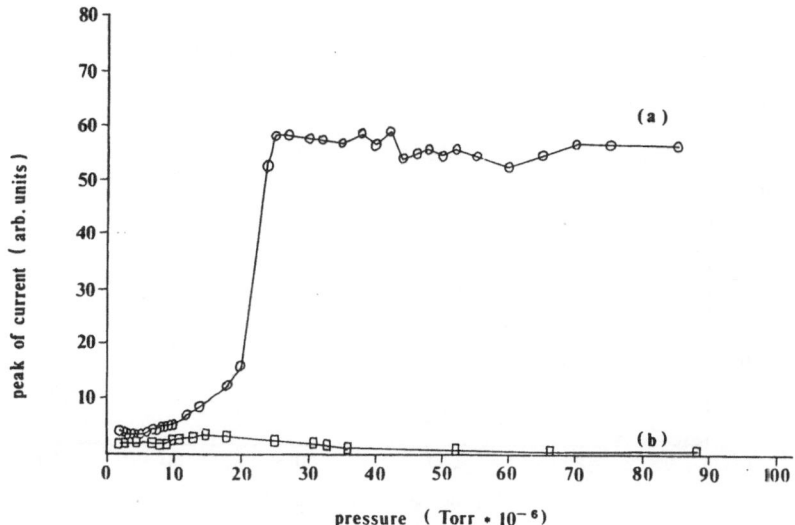

Fig. 1. Peak of the transmitted electron beam current as
a function of the overall gas pressure in the
vacuum chamber. Curve (a): gas injected into the
vacuum chamber through the needle injector;
curve (b): no needle injector. Note that, as
discussed in the text, a multiplicative scale
factor of 100±20 has to be used on the horizontal axis to obtain the actual pressure characterizing the points of curve (a).

The ion accelerating electric field (E=100 V/cm) is switched on as
soon as the electron pulse has left the ion source region (S). At the end
of the field free drift tube the ions are detected by a windowless focused
electron multiplier (EM). For the selected gas mixture the TOF's of the
different ionic species range between 4 µs and 5 µs. The overall time
resolution of the ion spectrometer is better than 1 ns.

The dependences of the electron current and total ion count rate on
the gas pressure, i.e. on the produced charge density, display the most
interesting features characterizing this experiment.

Figure 1a shows the behavior of the peak of the transmitted current
as a function of the mean value of the gas pressure in the chamber, when
the gas mixture is introduced in the vacuum chamber through the needle
injector. The axial current is insensitive to pressure variations up to
about 10^{-5} Torr, corresponding to 10^{-3} Torr just above the needle injector. Above this value the linear growth between $\cong 10^{-5}$ Torr and $\cong 2 \times 10^{-5}$
Torr is followed by a sudden dramatic increase, over a very small pressure
variation, that subsequently leads to a plateau region, where the peak
value is about ten times greater than the unperturbed one. A further
increase of pressure above $\cong 2.5 \times 10^{-5}$ Torr introduces only a small modulation around the maximum achieved value. The dependence of the total collected charge of the electron pulse current exhibits the same behavior.

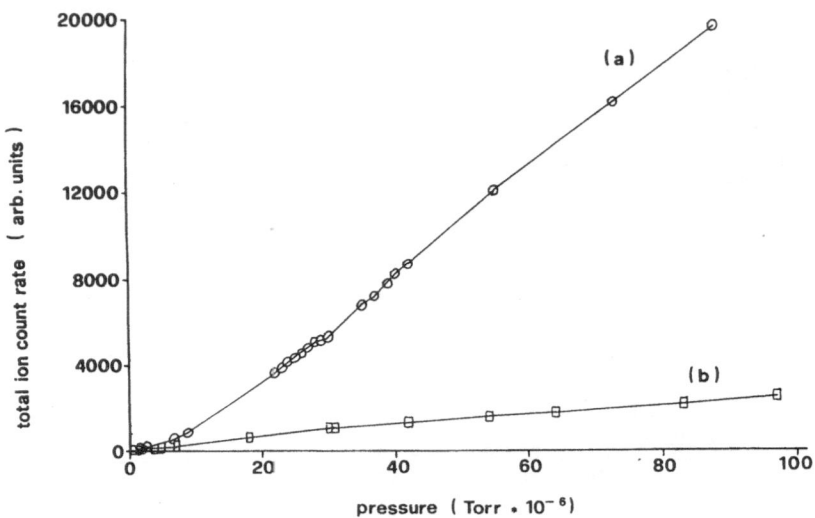

Fig. 2. Total ion count rate as a function of the gas pressure.
Cases (a) and (b) as in Figures 2 and 3.

In this case the maximum ratio between the total charge of the unperturbed current and the plateau value is about 30.

Figure 1b represents the experimental results obtained when the gas mixture is introduced in the vacuum chamber without a needle injector (uniform gas pressure). In this case the focussing effect is not observed. Recalling that the local pressure in the case of gas injected by the needle is 100 times the mean pressure, it is evident that the density of produced charges is strongly reduced. Moreover, in this case the dimension of the interaction zone is only determined by the cross section of the electron beam. This leads to a smoother charge density gradient than in the previous measurements.

The appearance of the focussing electric field depends on the rate of charge separation during the interaction time and, particularly, on the electron mobility, which is a function of the density gradient (see below). Hence, in the case of uniform gas pressure, the absence of focusing effects is ascribed not only the low level of produced charge density, but even more so to the strong decrease in its gradient.

Figure 2a shows the dependence of the total ion count rate on the gas pressure as observed when the gas mixture is injected through the needle. As a comparison, the same measurements, performed under conditions of uniform gas pressure, are represented by Figure 2b. Up to $\cong 10^{-5}$ Torr both curves overlap. As the pressure is increased, in the same region where the sudden growth of the electron current is observed, the ion dependence, although still linear, exhibits a relevant change in slope, whereas no modifications occur in the case of uniform pressure. A more detailed inspection of Figure 2 shows a sort of modulation of the slope in the

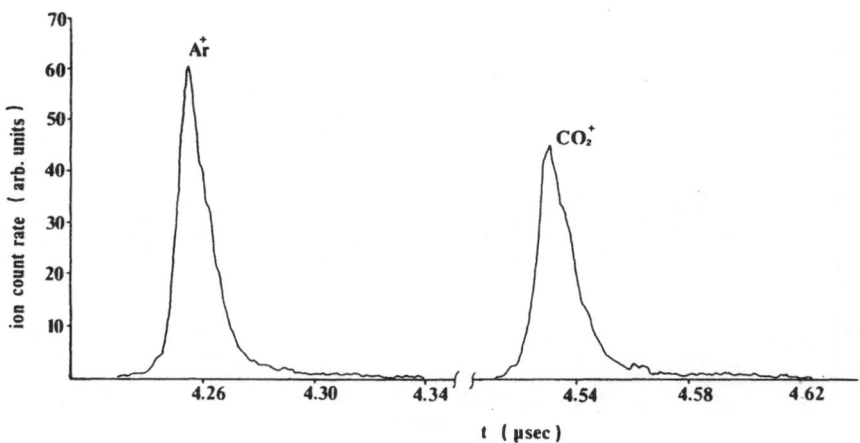

Fig. 3. TOF ion mass spectrum obtained with the gas injected through the needle and with a mean pressure of 3.8 x 10^{-6} Torr.

region between $\cong 10^{-5}$ Torr and $\cong 4 \times 10^{-5}$ Torr in correspondence to the modulation of the electron current.

The TOF spectrum of the ions also shows interesting differences if observed at different gas pressures. Whereas in the pressure range where no focusing effects occur the spectrum is constituted by single well defined peaks (Fig. 3), on the contrary, the sharpness of the peaks is reduced as the pressure increases. Moreover, relevant tails in the region of longer TOF's are observed (see Figure 4).

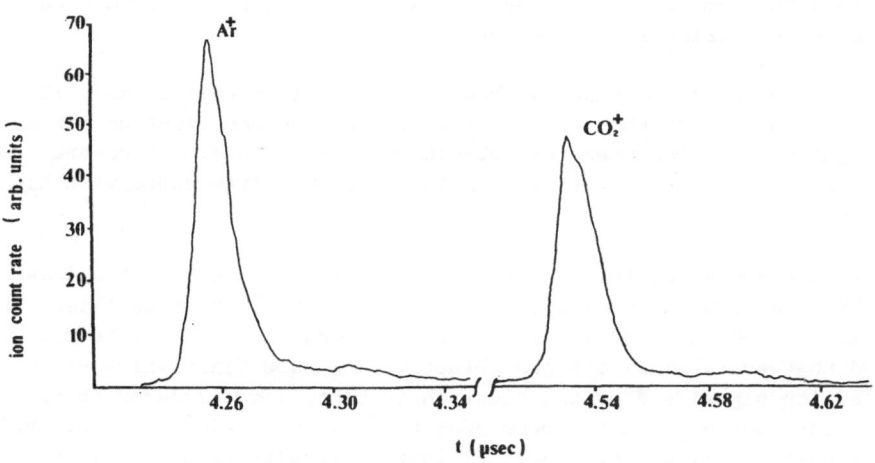

Fig. 4. TOF ion mass spectrum in the same conditions of Fig. 3 but with a mean gas pressure of 3.3 x 10^{-5} Torr.

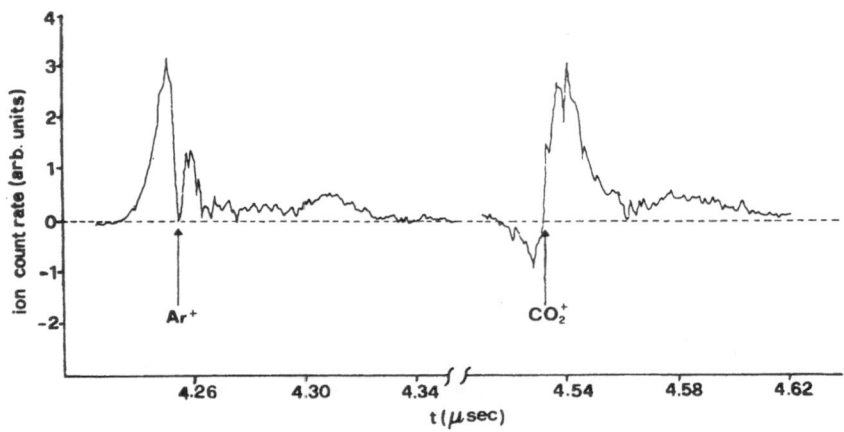

Fig. 5. TOF spectrum obtained by subtracting the spectrum of
Fig. 3 from that of Fig. 4. The vertical arrows indicate
the locations of the maxima of the Ar^+ and CO_2^+ peaks of
Figs. 3 and 4.

In order to better show these differences, Figure 5 represent what is
obtained by subtracting the spectrum of Figure 3 from that of Figure 4.
In Figure 5, the vertical arrows indicate the location of the maxima of Ar^+
and CO_2^+ peaks of Figures 3 and 4. The arrows correspond to zero counts
because both Ar^+ and CO_2^+ peaks have been normalized to the same values.
The spectrum of Figure 5 highlights the presence of both a relevant and
asymmetric broadening of the peaks, and of considerable tails in the
region of longer TOF's.

These last results support an interpretation based on collective
plasma effects due to the secondary produced charges. The presence in the
TOF ion spectrum of the tails of Figure 4 resembles those observed in Ref.
12, where a Na beam was ionized by a laser pulse and the ions were collected by an electric field of about 14 V/cm.

The MPI experiment in Na has been described in detail in Ref. 12.
Here, some features of the observed behavior of ion and electron yields
which highlight the influence of collective effects on detection are
briefly summarized with regard to similarities and differences with EI
experiments.

The dependences of ion yields and of electron current on the laser
intensity have been investigated for several three-photon transitions in
the region 539-600 nm, and at different beam densities. It has been
observed that every slope changes abruptly at a same fixed value of the
collected ion signal and of the electron current, i.e. produced charge
density, independently on the beam density (Fig. 6). Below that threshold,
every resonance follows the usual MPI laser intensity dependence. The
threshold appears at $\cong 10^{10}$ charges/cm^3 for ions, and at one order of
magnitude less for electrons.

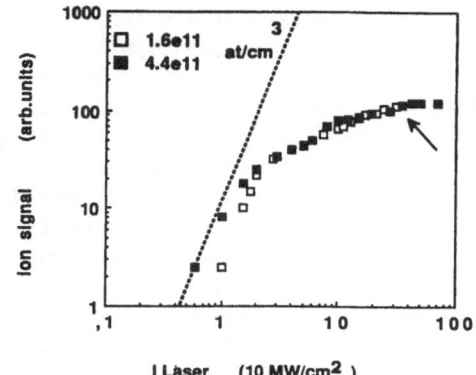

Fig. 6. Ion signal vs laser power for
the three-photon transition
$\lambda = 578.7$ nm in Na at beam densities of 1.6×10^{11} atoms/cm^3
and 4.4×10^{11} atoms/cm^3.
The arrow indicates the laser
intensity which corresponds
to total ionization of the Na
beam. (See Ref. 12).

Below the threshold, the time shape of electron and ion signal is
represented by a symmetric sharp peak, whose width depends on the dimension of the ionization region, and on the random motion of charges. The
time-of-flight corresponds to that calculated for uncoupled changes.
Around the threshold, electron and ion time shapes are modified, so that
the sharp peak is followed by a long tail extending towards longer time
delays. A further increase of the laser intensity leads to the appearance
of broad profile in the charge signal, whose maximum presents an arrival
time delayed in comparison to that expected for uncoupled particles. The
temporal profile of the electron signal contains a first peak, with delay
time in the ns range and associated with untrapped electrons, and a second
broad peak whose temporal evolution closely resembles the ionic ones. The
arrival time of the delayed electron peak is in the μs timescale typical
of ionic motion.

Whereas in the EI experiment collective effects increase the total
efficiency of ion collection, Figure 6 shows that in MPI electron-ion
interactions reduce the collection efficiency. In the next section, it
will be demonstrated that the ion focusing is caused by the excess of
negative charges due to the electron beam in the EI experiment, unlike the
MPI experiment where electrons and ions are produced at the same rate.

III. THEORETICAL APPROACH

As previously pointed out[11,13], the fluid approximations constitutes
a suitable approach to describe the time-space evolution of charges in

spite of the fact that the velocity distribution function is, in general, quite far from the equilibrium Maxwell distribution. In fact, the fluid scheme, derived from the Boltzmann transport equation as momenta of different order averaged over the velocity distribution function, holds true providing that the averaged particle velocity is equal to zero. Thus, an isotropic velocity distribution, regardless of its shape, is sufficient to make the fluid approach valid. The basic equations describe the time-space evolution of the fluid mean quantities (velocity, density, and energy) referred to the laboratory frame. It turns out that they are functions of coordinates and of time. The fluid velocity represents the common component of collective motion, whereas the temperature is related to the mean kinetic energy of random motion of the particles. In the case of isothermal expansion in vacuum, two partial differential equations, for fluid velocity and density, are sufficient to characterize the evolution, namely, motion and continuity equation[19]. For each species, the basic equations are given by

$$\frac{\partial \vec{v}_j}{\partial t} + (\vec{v}_j \nabla) \, \vec{v}_j = \frac{e_j}{m_j} (\vec{E}_0 + \vec{E}_r) - \frac{2E_j}{m_j} \nabla \ln (n_j) \tag{1}$$

$$\frac{\partial n_j}{\partial t} + \nabla \, n_j \, \vec{v}_j) = S_j \tag{2}$$

where $\vec{v}_j$, e_j, m_j, and n_j are the collective velocity, charge, mass and density of the selected species. E_0 represents the external electric field. The subscript j refers to the incident electrons (p) and to the secondary electrons and ions (j=e,i) produced by impact ionization, respectively. S_j represents the source term responsible for the growth of ionization ($S_p \overset{.}{=} 0$ for the incident electron beam). Since temperature is, in the EI and MPI experiments, a meaningless quantity in view of the very long electron-electron and electron-ion equipartition time[19], in Eq. (1) E_j is related to the mean energy of random motion in place of the temperature[11,13]. Thus in the case of incident electron, E_j is equal to the thermal spread (E_B) of the beam. In the case of secondary electrons and ions in EI experiment, E_j is equal to the mean kinetic energy of ejected electrons (E_{se}), and to the gas kinetic energy (E_{si}), respectively. In MPI experiments, E_j corresponds to the energy of ejected electrons given by $E_j = nh\nu - I$, where I is the ionization potential.

In Eq. (1) $\vec{E}_r$ is the self-generated electric field given by Poisson equation:

$$\nabla \vec{E}_r = 4\pi e (n_j - n_e - n_p) \tag{3}$$

Thus, Eqs. (1) and (2) for each species are coupled together by the internal field. Of course, n_p is zero in the MPI experiment.

Owing to the short ionization time (10 ns) and a very small interaction volume that implies strong density gradients, the complete system of equations cannot be simplified by using the usual plasma physics approximations, based on local neutrality and small time-space variation of macroscopic quantities (charge density, self-generated field, and so on)[19].

256

An approximate evaluation of the term depending on the density gradient in Eq. (1) shows that the dimensions of the interaction zone play a crucial role in the observed focusing phenomena, as confirmed by the experimental results of Sect. 2, and by comparison with those of Ref. 12. In fact, the electrons produced initially leave the interaction volume, whose width is d, as an effect of their random velocity in a character- istic time $\tau_{se} = (m/4E_{se})^{1/2}d$, the external field being zero during the interaction time. A positive internal field, due to the ions at rest, stops the decrease of electron density and reverses the electron motion. Therefore, incident and produced electrons are focused and their local density overcomes the ionic one, because of the excess of negative charges due to the incident electron beam, in such a way that ions too begin to be focused. The further ion motion in the external field is then affected by the preceding coupled interaction among incident electrons and produced electrons and ions.

First of all, the efficiency of the process depends on the time scale of electron expansion compared with the duration of the interaction, which in the present experiment is ≅10 ns. In the case of gas injected by a needle (d ≅ 0.01 cm), $E_{se} \cong 2$ eV, and thus $\tau_{se} \cong 0.1$ ns, whereas in presence of uniform gas pressure $\tau_{se} = 4$ ns, with a cross section of the electron beam of ≅0.4 cm. Moreover, the growth of the restoring ion field depends on the rate of produced charges, i.e. on the intensity of the incident current and the gas pressure. Finally, the net excess of negative charges, depend- ing on the intensity of the incident current, is responsible for the ion focusing. Since n_p is zero in MPI experiments, the internal field oscil- lates between zero and positive values. Thus, the positive internal field accelerates ions, and this leads to an increase of the interaction volume.

A quantitative description of the phenomena is simplified if the spatial profile of the species involved is represented by a Gaussian- shaped profile. It is demonstrated in Ref. 11 that the uncoupled motion of each species. i.e. with $E_r=0$, is still represented by a Gaussian-shaped profile, whose width is a function of time and whose maximum translates according to single-particle motion in an external field E_0. Thus, the one-dimensional time-space evolution of each density profile in presence of an uniform electric field E_0 along the x axis may be expressed as

$$n_j(x,t) = n_0 f(t)_j^{1/2} \exp\left[-f_j(t)(x-x_j^s)^2/d^2\right] \qquad (j=i,e,p) \qquad (4)$$

where d and n_0 represent the initial charge distribution width and density, respectively, while $x_j^s = \int v_j^s \, dt$, with $Dv_j^s = e_j E_0(t)/m_j$, represents the electron or ion single-particle motion in the external field, D being the total time derivative. The continuity equation provides the functional dpendence of the fluid mean velocities on n_j and, thus, on the $f_j(t)$ functions appearing in Eq.(4), namely $v_j = v_j^s + 0.5(x_j^s - x) \cdot |Dln(f_j(t))|$. The functions $f_j(t)$ satisfy the initial conditions $f_j(t=0)=1$, and $Df_j(t)|_{t=0}=0$. A second order differential equation describing the $f_j(t)$ time evolution is derived from the equation of motion by replacing the $f_j(t)$ functional dependencies for both mean velocity and n_j, and by equating the terms with the same power dependence on the spatial coordi- nates. Owing to the independence of motion along the axes, the three-dimensional motion is represented by a product of three Gaussian profiles with different initial widths, depending on the symmetry of the charge production. Thus, the time evolution of $f_j(t)$ along any axes depends on the initial width.

In general, the inclusion of the internal field E_r and of the source terms S_j makes the previous scheme of solution invalid. Nevertheless, under suitable conditions both terms can be reduced to a form such that the basic scheme of solution still holds true.

The Poisson equation is linearized around the coordinates of the single-particle motion for the electrons and ions, i.e. around the top of the Gaussian profiles. As an example of the technique, in the one-dimensional case and for two generic electron and ion species, integrating between $-\infty$ and x, and introducing the variable change $\varepsilon_j = f_j^{1/2}(x_j^s - x)/d$, where the index j refers to both the species, the internal field is given by

$$E_r(x,t) = 4\pi edn_0 \int_{\varepsilon_i}^{\varepsilon_e} e^{-\varepsilon^2} d\varepsilon, \qquad (5)$$

where the subscripts i and e refer to ions and electrons, respectively.

By series integration, retaining the first term only, one immediately obtains

$$E_r(x,t) = 4\pi en_0 \left[f_i^{1/2}(x-x_i^s) - f_e^{1/2}(x-x_e^s) \right]. \qquad (6)$$

Eq. (6) is valid insofar as the second term of the series development is negligible. This allows one to obtain the spatial interval of validity of the previous solution from the condition

$$\varepsilon_e - \varepsilon_i \geq (\varepsilon_e^3 - \varepsilon_i^3)/3 \qquad (7)$$

In the form of Eq. (6), the internal field can be introduced into the basic system of Eqs. (1) and (2), and by equating terms with the same power of x, new differential equations for $f_j(t)$ and $x_j^s(t)$ are immediately obtained.

When condition of Eq. (7) is not met, the motion is still represented by a Gaussian profile freely expanding, i.e. with $E_r = 0$, weighed by the total number of particles able to overcome the internal field, namely

$$n_f = k(t)f(t)^{1/2}\exp(-f(t)(x/d)^2).$$

k(t) is given by mass conservation, i.e.

$$n_0 2 \int_0^{\varepsilon^*} e^{-\varepsilon^2} d\varepsilon + k(t)2 \int_{\varepsilon}^{\infty} e^{-\varepsilon^2} d\varepsilon = \sqrt{\pi} \, dn_0, \qquad (8)$$

where ε^* is a value that satisfies condition (7). The first integral is calculated by introducing the value of f(t) given by the coupled system, whereas in the second integral f(t) is the solution with $E_r = 0$.

The boundary between coupled and uncoupled zones can be obtained only by an experimental analysis of the ratio between trapped and untrapped particles as in Ref. 12.

As in the case of freely expanding motion, extension to three dimension is straightforward, and in that case the internal field depends on the square root of the product of three $f_j(t)$ functions.

The production term S_j can be introduced into preceding analysis at a degree of approximation of the same order of that used in linearizing Poisson equation[12]. In fact, it is sufficient to replace n_0 in Eq. (4) by the solution $g(t)$ of the rate equation:

$$\frac{dg(t)}{dt} = <a> \, n_p \, (N - g(t)) \tag{9}$$

In Eq. (9) N represents the unperturbed gas density, n_p the electron beam density and $<a>$ (cm^3/sec) the ionization cross section averaged over the velocity distribution funtion. It has been also assumed that only ions and molecules in the fundamental state are present in the gas. In the case of the MPI experiment, Eq. (9) is replaced by similar rate equation containing the laser intensity at the power of the selected MPI transition.

The geometrical characteristics of the EI experiment lead to an important simplification in the number of differential equations to be solved. The incident beam is taken to be cylindrically symmetrical around the propagation axis (x), and is described by radial and longitudinal Gaussian profiles, whose widths are d_r and d_x, respectively. It is a reasonable approximation to assune spherical expansion of the produced charges. Thus, the complete scheme includes a system of four differential equations, of which two refer to the longitudinal and radial f(t) functions of the beam (respectively $f_{px}(t)$ and $f_{pr}(t)$), and the other two to both secondary ion and electron f(t) functions ($f_{si}(t)$ and $f_{se}(t)$), respectively.

The differential equations for the $f_j(t)$ functions, during the interactions time ($E_0=0$) are given by

$$D^2 f_j = \frac{3(Df_j)^2}{2 \, f_j} - 2(f_j/\tau_j)^2 + C(g(t), f_i, f_e, f_{px}, f_{pr}) \tag{10}$$

where τ_j is the characteristic time constant of free expansion due to the density gradient of the involved species, respectively given by

$$\tau_{px}^2 = \frac{m_e \, d_x^2}{4E_B} \, , \quad \tau_{pr}^2 = \frac{m_e \, d_r^2}{4E_B} \, , \quad \tau_{sj}^2 = \frac{m_e \, d_s^2}{4E_{sj}} \tag{11}$$

where E_B is the energy spread of the electron beam, and d_x and d_r its longitudinal and radial widths, respectively, and j=i,e for secondary charges. E_{se} represents the mean energy of the secondary electrons, produced in an almost spherical interaction volume of radius d_s, whereas E_{si} is given by the gas temperature. The coupling term $C(g(t), f_j, f_{px}, f_{pr})$, due to the internal field, is for each species given by

$$\frac{2}{3} \left(\frac{4\pi e^2}{m_e}\right) \left[(Ng(t)f_{si}^{3/2} - (Ng(t)f_{se}^{3/2} + n_p f_{px}^{1/2} f_{pr})]f_{px},$$

$$\frac{2}{3} \left(\frac{4\pi e^2}{m_e}\right) \left[(Ng(t)f_{si}^{3/2} - (Ng(t)f_{se}^{3/2} + n_p f_{px}^{1/2} f_{pr})]f_{pr},$$

$$\pm \frac{2}{3} \left(\frac{4\pi e^2}{m_e}\right) \left[(Ng(t)f_{si}^{3/2} - (Ng(t)f_{se}^{3/2} + n_p f_{px}^{1/2} f_{pr})]f_{sj}. \tag{12}$$

The last terms refers to secondary charges and the upper (plus) sign to the electrons. The initial conditions are of course $f_j(t)=0$ and $Df_j(t)|_{t=0}=0$. The coupling terms given by Eqs. (12) are effective during an interaction time that depends on the duration of the electron beam pulse and on its velocity v_0, namely $t_{int}=d_x/(f_{px}^{1/2} v_0)$. If the internal field is negligible ($f_{px}=1$), that time is equal to the unperturbed pulse duration $\cong 10$ns.

For $t>t_{int}$ the coupling therm disappears in the equations which describe the freely expanding electron beam, but the values of the $f(t)$ functions and its derivatives are as obtained at the end of the mutual interaction. In the equations for secondary charges, at the end of the interaction, the term depending on n_p vanishes, the function $g(t)$ assumes the constant value $g(t_{int})$ and two single-particle-like equations must be added to take into account the switching on of the the external field, namely

$$D^2 x_j^s = \frac{e_j}{m_j} E_0 \pm \frac{1}{3} \omega_{pj}^2 f_j^{3/2} (x_i^s - x_e^s) \tag{13}$$

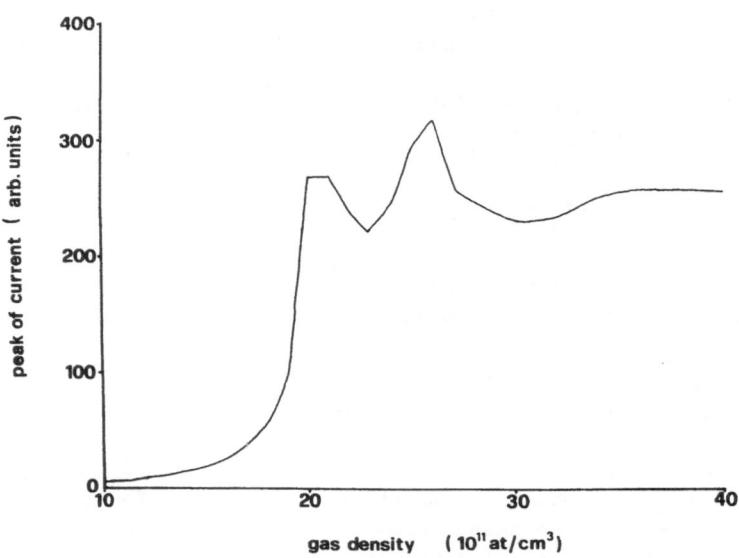

Fig. 7. Theoretical calculation of the dependence
of the electron current peak on gas density.

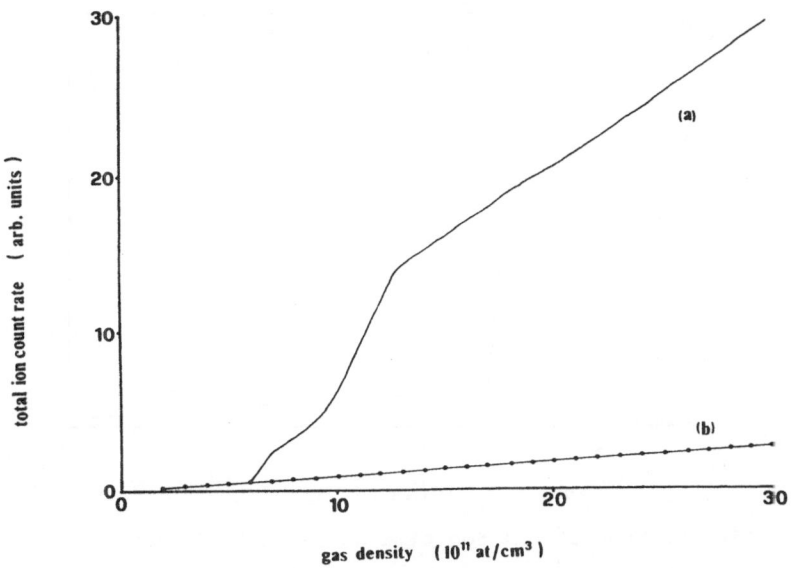

Fig. 8. Theoretical dependence of the total ion count rate
as a function of the gas density in the case of coupled
(curve (a)) and uncoupled (curve (b)) expansion of the
charge species.

where the upper sign refers to electrons; $\omega_{pj}=4\pi e^2 N g(t_{int})/m_j$ is the plasma frequency. In Eq. (13) the index j denotes that if x_j^s refers to ions, f_j refers to electrons, and vice versa. The initial conditions are $x_j^s(0)=0$ and $Dx_j^s(t)|_{t=0}=0$. The temporal origin of the ion motion is now chosen so as to correspond to the instant of the switching on of the field. The new initial conditions of integration of Eq. (11) for $f_j(t)$ and $Df_j(t)$ are those determined at the end of the interaction with the incident beam.

Fig. 7 shows, as a function of the gas density, the behavior of the peak of the electron current in typical experimental conditions, that is E_B = 0.1 eV, E_{se} = 2 eV, E_{si} = 0.1 eV and d_x = 2 cm, d_r = 0.1 cm, d_s = 0.01 cm. We have also assumed an ionization rate ($<a>n_p$) of 10^4 sec^{-1} as approximately determined from the measurements performed at uniform gas pressure (see Figure 2b) and a density of incident electrons of 10^9 electrons/cm^3. The dependence of the total ion count rate in an external field of $\cong 100$ V/cm (Sect. II) on the gas density is shown in Figure 8 in the case of coupled (curve a), and uncoupled (curve b) expansion. Figure 9 shows a typical theoretical ion time-shape.

A simple estimation of the threshold of the phenomena can be carried out by comparing the coupling terms of Eqs. 12 with the respective time scales τ_j, at the end of the beam pulse, and by assuming that the f_e function of secondary electron is negligible owing to the rapid expansion ($\tau_{se}\cong0.1$ ns), whereas f_j is close to one (ions are approximately at rest). In Eqs. 12, referred to the incident beam, the positive term inducing

261

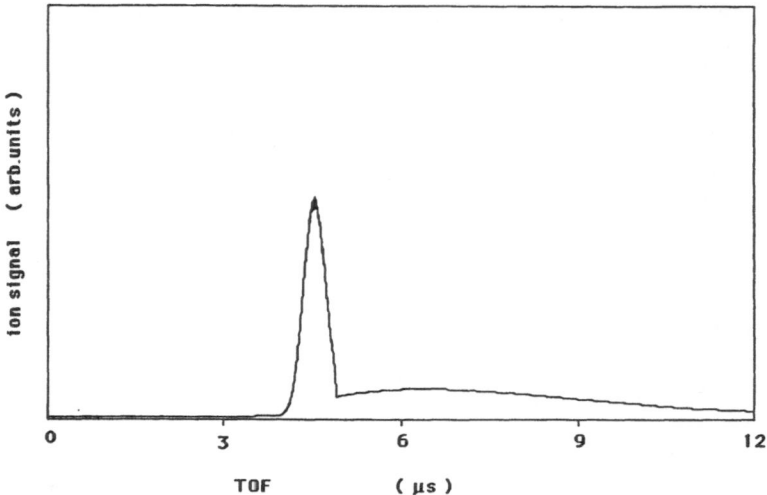

Fig. 9. Theoretical ion time shape at 10^{-4} Torr.

growth of the f_{px}, f_{pr} functions (focussing) is given by $(8\pi e^2/3m_e)$ · $Ng(t_{int}) = (8\pi e^2/3m_e)\bar{N}\langle a\rangle n_p t_{int}$; t_{int}, near the threshold of the phenomenon, is equal to the duration of the beam pulse (10 ns), and $g(t)$ is simply given by $g(t)=\langle a\rangle n_p t$, owing to the low ionization rate. The negative terms depend on the electrostatic repulsion $((8\pi e^2/3m_e)n_p$, with $f_{px}=f_{pr}=1)$ and on the random spread (τ_{px} and τ_{pr}). Thus, comparing positive and negative terms, the threshold condition is given by

$$N\langle a\rangle n_p t_{int} \geq \frac{3E_B}{2\pi e^2 d_j^2} + n_p \qquad (14)$$

where E_B represents the energy spread of the beam, and the index j its radial or longitudinal width. Upon introducing the foregoing approxima-tions, Eqs. (14) leads to a pressure value, for the threshold, of $\cong 3\times 10^{-4}$ Torr. Considering that the actual pressure over the needle is about 100 times the background, this value is in a good agreement with the measured one ($\cong 10^{-3}$ Torr see Figures 1 and 2). It turns out from Eq. 14 that, in the case of negligible beam thermal spread (first term in the right side $\langle n_p\rangle$), as in the present experiment, the pressure threshold allows one to esti-mate the electron impact ionization rate, i.e. $N \geq 1/\langle a\rangle t_0$.

In Eq. (10) relative to the ions, the terms responsible for the focusing are related to the density of primary and secondary electrons. During the interaction time and near or above the threshold, the density of secondary electrons decreases or oscillates around the value of the ion density, whereas the density of primary electrons suddenly increases. When $n_p f_{px}^{1/2} f_{pr}$ becomes greater than $Ng(t)$ (the ion thermal spread being negligible during the interaction time), the ion f_{si} function increases as an effect of the positive coupling term. Therefore, ion focusing must occur in the same pressure range corresponding to the sudden growth of the incident beam density, i.e. transmitted current.

262

The estimation of the charge density threshold in the case of MPI, can be obtained from Eq. (12) for secondary charges with $n_p = 0$, and this leads to

$$n_e \geq \frac{4E_{se}}{4\pi e^2 d^2},\tag{15}$$

Eqs. (14) and (15) can be also obtained from the neutrality condition implying a Debye length smaller than the plasma characteristic length[19]. Eq. (15) gives $n_e \geq 3.10^8$ electrons/cm , in agreement with the results of Refs. 12 and 13. The observed ion threshold in Ref. 12 corresponds to a produced charge density roughly 10 times larger than the electron density. The expansion of the ion profile is enhanced by energy transfer from electrons to ions. Thus, terms in Eq. (10) related to the ionic plasma frequency and to the square derivative play a dominant role. Therefore, the threshold value for ions can be determined only by numerical calculations.

IV. CONCLUSIONS

The EI experiment confirms the influence of collective effects in a range of relatively low charge density (10^8-10^{10} charges/cm^3), as observed in the MPI experiment of Ref. 12. Common features have been observed in the ion time shape, i.e. a spread of the initial profile, and delay of the time of flight, whose importance depends on the experimental conditions, as the value of the external electric field and the produced charge density. In fact, in the EI experiment, with an electric field of 100 V/cm and a charge density of $\cong 10^9$ charges/cm^3, only a long tail in the TOF ion distribution is observed, whereas in the MPI experiment, with an electric field of $\cong 14$ V/cm and a charge density of $\cong 10^{10}$ charges/cm^3, broadening of the peak and TOF delay were evident[12].

If the overall purpose is the collection of selectively ionized species, as in isotope separation experiments or in trace analysis, the spread of the ion profile during the coupling phase, could make it difficult to separate atoms and/or molecules, whose weight differs by only a few mass units. In fact, in this case, the delayed ions of the most abundant isotope can arrive at a time that corresponds, on the scale of the spectometer, to the position of the isotope of interest, or, at the very least can constitute a sort of diffused background reducing the resolution of the spectrometer.

Unlike the MPI experiment, a marked improvement of ion collection efficiency is observed in the EI experiment due to ion focusing. Ion focusing occurs during the interaction time with the electron beam when the external field is switched off, whereas spread of the profile occurs during the application of the external field. Thus, in principle, the experimental parameters can be chosen in such a way as to maintain ion focusing without spread of the TOF distribution. In such a case, collective effects should improve collection efficiency. Moreover, the same effect on the incident beam could be used to increase collimation, and hence, intensity of electronic beams.

Electron and ion focusing is a transient phenomenon. In fact, the amplitude of the electronic oscillations is reduced as the ion density increases. As ion density reaches the maximum value, electrons oscillate close to the ions, and a steady state is established in which the neutrality condition is fulfilled. Thus, the EI experiment provides a suitable opportunity to study nonlinear phenomena related to the time evolution of nonequilibrium plasmas.

REFERENCES

1. J. A. Paisner, _Appl. Phys._ B 46, 253 (1988)
2. Yu. A. Kudryavtsev, V. V. Petrunin, V. M. Sitkin and V. S. Lethokov, _Appl. Phys._ B 48, 93 (1989)
3. For a review of the influence of "space charge" phenomena in different field of investigation and mostly application see "Resonance Ionization Spectroscopy-III", G. S. Hurst ed., Institute of Physics Conference Series n. 84, Institute of Physics, Bristol, (1987), and "Resonance Ionization Spectroscopy-IV", T. B. Lucatorto and J. E. Parks eds., Institute of Physics Conference Series n. 94, Institute of Physics, Bristol (1988)
4. G. I. Bekov and V. S. Lethokov, in: "Laser Analytical Spectrochemistry", V. S. Lethokov, ed., Adam Hilger, Bristol (1986). See also papers of the same authors in Ref. 3 and references therein. Also in this book.
5. A. M. Malvezzi, H. Kurz and N. Bloembergen, Mat. Res. Soc. Symp. Proc., 35, 2407 (1985). See also papers of T. B. Lucatorto and G. S. Hurst in Ref. 3
6. M. Crance, _J. Phys._ B 19, L267 (1986)
7. G. Petite, P. Agostini and F. Yeargeau, _J. Opt. Soc. Am._ B 4, 765 (1987)
8. L. A. Lompre', A. L'Huillier, G. Mainfray and C. Manus, _J. Opt. Soc. Am._ B 2, 1906 (1985)
9. T. J. McIlrath, P. H. Bucksbaum, R. R. Freeman and M. Bashkansky, _Phys. Rev._ A 35, 4611 (1987)
10. B. Carre', F. Roussel, G. Spiess, J. M. Bizeau, P. Gerard and F. Wuilleumier, _Z. Phys._ D 1, 79 (1986)
11 F. Giammanco, _Phys. Rev._ A 36, 5658 (1987)
12. F. Giammanco, _Phys. Rev._ A 40, 5160 (1989)
13. F. Giammanco, _Phys. Rev._ A 40, 5171 (1989)
14. G. I. Guseva and M. A. Zav'yalov, _Sov. J. Plasma Phys._, 9, 4, 445 (1983)
15. V. P. Kovalenko and S. F. Fastovets, _Sov. J. Plasma Phys._, 9, 5, 562 (1983)
16. M. Armenante, V. Santoro, N. Spinelli and F. Vanoli, Int. J. Mass Spectrom. and Ion Proc., 64, 265 (1985)
17. G. Arena, M. Armenante, S. Salerni and N. Spinelli, Rapid Communic. in Mass Spectrom., 2, 35 (1988)

18. G. Giammanco, G. Arena, R. Bruzzese and N. Spinelli, Phys. Rev. A 41, 2144 (1990)
19. L. Apitzer, Physics of Fully ionized gases, Interscience, New York (1956)

PRINCIPLES OF COMPLEX INTERFEROMETRY

M. Kalal

Technical University of Prague
Faculty of Nuclear Sciences and Physical Engineering
Department of Physical Electronics
Prague, Czechoslovakia

I. INTRODUCTION

Interferometry in its classical form is used for phase imaging of optically transparent objects. Up to now many different techniques for phase shift reconstruction from interferograms have been developed: from the most primitive and the less accurate use of the ruler through the extremes of fringe seeking computer algorithms up to the most sophisticated and very accurate analysis based on the fast Fourier transform (FFT)[1].

Because this last mentioned technique makes possible a simultaneous calibration of the recording medium, it became immediately apparent that in the case of high quality beams (i.e. both the probe and the reference beam) not only the phase shift of the probe beam (encoded as a frequency modulation) but also the amplitude of the probe beam (encoded as an amplitude modulation) should have a good chance for reconstruction from just one interferogram. This approach could bring substantial advantages over the usual method of recording two sets of data separately, as it is very difficult to spatially superimpose these two sets of data during analysis.

This idea of two sets of data recorded in one interferogram led to the development of the phase-amplitude imaging technique, which was successfully tested on simulated data[2]. The first real application of this technique was the fully automated analysis of megagauss magnetic field measurements in laser-produced plasmas. During the analysis of experimentally obtained interferograms, however, one more phenomenon - fringe blurring caused by th ephase shift changes during an exposure time- was found to have a detrimental effect on the accuracy of reconstructed data[3]. On the other hand, this initially rather unpleasant surprise revealed the fact that the FFT analysis technique is so sensitive that it can be used, in principle, for the reconstruction of up to three independent sets of data encoded into single interferogram: the phase shift, the amplitude and the phase shift time derivative.

This generalization of classical interferometry we shall, from now on, call "complex interferometry". Its mathematical description in the general case, as well as several particular examples, will be given in the subsequent paragraphs.

II. COMPLEX INTERFEROMETRY

II.1. General Description

If a coherent probe beam with its amplitude varying in time passes through an object that modifies both its phase and amplitude, the resulting beam may be described by

$$\psi_p(\vec{r},t) = a(\vec{r},t)f(t)\exp\left[ik(\vec{r} + i\phi\vec{r},t)\right]. \tag{1}$$

After this interferes with a reference beam

$$\psi_r(\vec{r},t) = a_0 f(t)\exp(i\vec{k}\cdot\vec{r}) \tag{2}$$

and choosing the (y,z) plane as the plane of interference, a time dependent intensity pattern described by

$$i(y,z,t) = a_0^2 f^2(t) + a^2(y,z,t)f^2(t) +$$
$$+ 2a_0 a(y,z,t)\cos\left[2\pi(\omega_0 y + \nu_0 z) + \varphi(y,z,t)\right] f^2(t) \tag{3}$$

will result, where the relations between two sets of variables $\vec{k}_p$, $\vec{k}_r$, $\phi(\vec{r},t)$ and ω_0, ν_0, $\varphi(y,z,t)$ are straightforward. In this way, we have encoded both the amplitude and the phase changes of the probe beam introduced by the objects. The original temporal profiles of the amplitudes of both beams are denoted by $f(t)$ with the following normalization:

$$\int_{-\infty}^{+\infty} f^2(t)dt = 1. \tag{4}$$

Up to now, the expression (3) in its general form has not been solved. In the following paragraphs we shall describe the most important particular cases where this solution can be found.

II.2. Stationary Phase-Amplitude Objects

In the case of stationary phase-amplitude objects, integrating the intensity pattern (3) in time, the following expression will result:

$$i(y,z) = a_0^2 + a^2(y,z) +$$
$$+ 2a_0 a(y,z) \cos\left[2\pi(\omega_0 y + \nu_0 z) + \varphi(y,z)\right]. \tag{5}$$

A technique for accurately recovering both the phase shift, $\varphi(y,z)$, and the amplitude, $a(y,z)$, in this particular case has been proposed by

Kalal et al.[2]. To facilitate further comparisons we shall now briefly describe this procedure.

We may rewrite Eq.(5) as

$$i(y,z) = b(y,z) +$$

$$+ v(y,z)\exp[2\pi i(\omega_0 y + \nu_0 z)]+$$

$$+ v^*(y,z)\exp[-2\pi i(\omega_0 y + \nu_0 z)], \qquad (6)$$

where

$$b(y,x) = a_0^2 + a^2(y,z), \qquad (7)$$

$$v(y,z) = a_0 a(y,z)\exp[i\phi(y,z)]. \qquad (8)$$

The Fourier transform of Eq.(6) is easily shown to be

$$I(\omega,\nu) = B(\omega,\nu) +$$

$$+ V(\omega - \omega_0, \nu - \nu_0) + V^*(\omega + \omega_0, \nu + \nu_0), \qquad (9)$$

where the uppercase letters denote the Fourier transforms of the functions corresponding to the lowercase letters.

Thus, provided the bandwidth of $B(\omega,\nu)$ is sufficiently small and the fringe spatial frequency $f_0 = (\omega_0^2 + \nu_0^2)^{1/2}$ is sufficiently high, ther is negligible overlap between the shifted functions $V(\omega - \omega_0, \nu - \nu_0)$, $V^*(\omega + \omega_0, \nu + \nu_0)$ and $B(\omega,\nu)$. Here the terms V and V^* represent the fringe visibility function and contain the phase and amplitude information heterodyned away from the background signal B.

By shifting one of the sidelobes, V or V^*, back to zero frequency and windowing to exclude all information from the other sidelobe and from the background, and then retransforming into the (y,z) plane, the phase shift function $\phi(y,z)$ can be recovered from the imaginary part of the complex logarithm of the reconstructed function v(y,z) through the identity

$$\varphi(y,z) = \text{Im}[\log v(y,z)]. \qquad (10)$$

The fringe amplitude is obtained from the modulus of the function v(y,z) as below:

$$a(y,z) = |v(y,z)|/a_0. \qquad (11)$$

This technique was successfully tested on simulated data.

II.3. Non-stationary Phase Objects

In this paragraph we shall focus our attention to the very common case of non-stationary phase objects. Our aim will be to analyze this particular case using the approach of complex interferometry, and to compare our results with those obtained by classical interferometry.

The basic expression for the time dependent intensity pattern relevant to this problem can easily obtained from Eq.(3), replacing $a(y,z,t)$ by a_0 throughout:

$$i(y,z,t) = 2a_0^2 \; f^2(t) +$$

$$+ \; 2a_0^2 \; \cos \left[2\pi(\omega_0 y + \nu_0 z) + \varphi(y,z,t) \right] \; f^2(t). \tag{12}$$

The approach of classical interferometry is to neglect changes in the phase shift $\varphi(y,z,t)$ during the recording time, replacing it formally by some (initially unknown) time-independent function $\varphi_c(y,z)$, (see Ref. 1). Then following the procedure described in the previous paragraph, we arrive directly at the equivalent of Eq.(10):

$$\varphi_c(y,z) = \mathrm{Im}\left[\log v(y,z) \right]. \tag{13}$$

Let us now analyze Eq.(12) using the more consistent approach of complex interferometry, i.e. taking into account not only time changes of the phase shift during the probing pulse, but also considering effects of the shape of this pulse itself (e.g. symmetrical versus asymmetrical).

The time-dependent fringe visibility function $v(y,z,t)$ can be written in this case in the form analogous to Eq.(8)::

$$v(y,z,t) = a_0^2 \exp\left[i\varphi(y,z,t) \right] f^2(t). \tag{14}$$

For the purpose of a subsequent integration the time dependent phase shift $\varphi(y,z,t)$ can be approximated by a linear expansion in terms of a time invariant part $\varphi_0(y,z)$ and its time derivative $\varphi_c'(y,z)$, this being a reasonable approximation in the majority of practical cases:

$$\varphi(y,z,t) = \varphi(y,z,0) + t \; \frac{d\varphi(x,z,t)}{dt} \bigg|_{t=0} =$$

$$= \varphi_0(y,z) + t \; \varphi_0'(y,z). \tag{15}$$

Integrating Eq.(14) in time, after the substitution of Eq.(15) for $\varphi(y,z,t)$, the total visibility $v(y,z)$ has the following form:

$$v(y,z) = a_0^2 \exp\left[i \; \varphi_0(y,z) \right] q(y,z). \tag{16}$$

Here

$$q(y,z) = \int_{-\infty}^{+\infty} f^2(t) \exp\left[it \; \varphi_0'(y,z) \right] dt \tag{17}$$

becomes a modifying function with generally complex values, which can be rewritten in the form

270

$$q(y,z) = \exp|i\ \Phi_m(y,z)|. \tag{18}$$

In the classical interferometry this modifying phase $\Phi_m(y,z)$ represents a systematic error of the reconstruction. This error decreases with shortening of the probing pulse and completely vanishes in the case of a symmetric pulse about $t = 0$ when $\Phi_0(y,z) = \Phi_c(y,z)$.

In the complex interferometry approach the modifying function $q(v,z)$ can be used to recover the phase shift time derivative $\Phi_0'(y,z)$, provided that the shape of the probing pulse is known and symmetrical. In such a case the modifying function $q(y,z)$ becomes a function with real values only. Thus, the phase shift $\Phi_0(y,z)$ can be recovered the usual way

$$\Phi_0(y,z) = \text{Im}[\log v(y,z)] \tag{19}$$

and for the modifying function $q(y,z)$ the following expression holds:

$$q(y,z) = [v(y,z)]/a_0^2. \tag{20}$$

Now, considering the most typical case, when the shape of the probing pulse is monotically increasing and decreasing, the function

$$q(x) = \int_{-\infty}^{+\infty} f^2(t) \cos xt\ dt \tag{21}$$

is also a monotonic function of x. Hence, the inverse function q^{-1} exists (either in numerical of analytical form) and can be used to recover the phase shift time derivative:

$$\Phi_0'(y,z) = q^{-1}[q(y,z)]. \tag{22}$$

As an example, let us assume a typical probing pulse having a gaussian profile:

$$f^2(t) = (\tau\sqrt{\pi})^{-1} \exp(-t^2/\tau^2). \tag{23}$$

Then

$$q(x) = \exp(-x^2\tau^2/4) \tag{24}$$

and therefore

$$\Phi_0'(y,z) = 2\{-\ln[q(y,z)]\}^{1/2}/\tau. \tag{25}$$

This phase shift time derivative $\Phi_0'(y,z)$ contains one additional piece of information about the phase object under consideration which can be used, together with the phase shift information, to recover the velocity of this phase object.

In this paragraph's last brief remark let us specify the duration of the longest usable probing pulse. It is clear that for the purpose of any meaningful reconstruction at least some minimal contrast of fringes must be retained. This condition will be fulfilled provided that the change of

271

the phase shift during the probing pulse duration Δt is less then one fringe. This can be expressed mathematical as

$$\Delta t < 2\pi/\varphi_0'(y,z). \tag{26}$$

II.4. Non-stationary Phase-Amplitude Objects

This most general case can be completely solved only under following conditions: (i) the probing pulse is symmetrical about $t = 0$, (ii) the shape of the probing pulse is known, (iii) changes in the amplitude $a(y,z,t)$ during the probing pulse can be neglected, and (iv) the phase shift $\varphi(y,z,t)$ can be approximated by the linear expansion (15). Then the equivalent of Eq.(3) has the form:

$$i(y,z,t) = a_0^2 f^2(t) + a^2(y,z)\ f^2(t) +$$
$$+ 2a_0 a(y,z)\ \cos\left[2\pi(\omega_0 y + \nu_0 z) + \varphi(y,z,t)\right] f^2(t). \tag{27}$$

Integrating this equation in time we arrive at the expression (7) for $b(y,z)$ and at the equivalent of the expression (16) for $v(y,z)$:

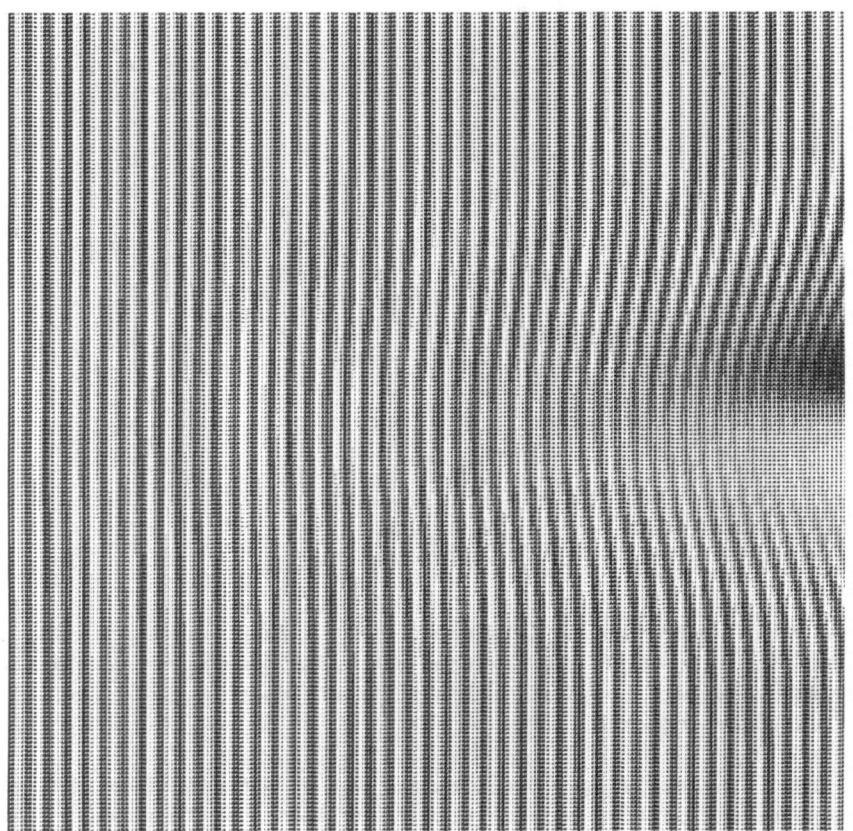

Fig. 1. Simulated complex interferogram.

$$v(y,z) = a_0 a(y,z) \exp[i\,\psi_0(y,z)]q(y,z). \tag{28}$$

It is then clear that the phase shift $\psi_0(y,z)$ can be recovered in the usual way using Eq.(19). For the amplitude $a(y,z)$, however, Eq.(28) cannot be used directly due to the presence of the modifying function $q(y,z)$ (see the previous paragraph). Instead, we must use Eq.(7) which yields:

$$a(y,z) = \left[b(y,z) - a_0^2\right]^{1/2}. \tag{29}$$

The remaining information $\psi_0'(y,z)$ can now be recovered from the expression

$$q(y,z) = |v(y,z)|/a_0 a(y,z) \tag{30}$$

followed by the procedure described in the previous paragraph. This approach has already been used in practice.

To illustrate this approach, a simulated complex interferogram relevant to the situation discussed in this paragraph is shown in Figure 1.

III. CONCLUSIONS

In this chapter the principles of a very powerful interferometric technique allowing the recording and reconstruction of up to three sets of data from just one interferogram have been described in detail. This technique provides highly accurate results and permits automatic film calibration. In combination with CCD cameras, this technique is well suited for fully automated on-line diagnostics and measurements.

REFERENCES

1. K. A. Nugent, Appl. Opt., 24:3101 (1985)
2. M. Kalal, K. A. Nugent and B. Luther-Davies, Appl. Opt., 26:1674 (1987)
3. M. Kalal, K. A. Nugent and B. Luther-Davies, J. Appl. Phys., 64:3845 (1988)

ADAPTIVE OPTICS IN THE FAR IR AND SUBMILLIMETER RANGES

A. B. Shvartsburg

Central Design Bureau for Unique Instrumentation
USSR Academy of Sciences
Moscow, USSR

I. INTRODUCTION

A new method for high speed temporal modulation of infrared wave beams, reflected from semiconductor thin films containing a plasma of free carriers, is proposed. The physical foundations of such a method, based on the non-linear and bistability properties of free carrier plasmas in these films, show the possibility of developing high-speed tunable modulators and polarizers. The bistability conditions are tunable by means of magnetic field. The reflection of a continuous wave beam from a film with tunable reflection coefficient leads to pulse shaping, the pulse duration being as small as 10^{-9} sec.

This chapter is devoted to a new possibility for high speed tunable modulation of amplitude, phase and polarization of far infra-red and sub-millimeter electromagnetic waves. Such a possibility is based on radiation reflection from semiconductor free carrier plasmas, the reflection coef-ficient may be tuned due to carrier heating, the incident wave frequency ω, the carrier's Langmuir frequency Ω_L and their frequency of collisions ν obeying the condition

$$\omega \cong \Omega_L \cong \nu \tag{1}$$

Unlike the well known effect of radiowave cross-modulation, resulting[1] from thermal variation of the imaginary part of the dielectric permit-tivity ε, due to carrier heating and frequency ν dependent upon the carrier temperature $\nu = \nu(T_e)$, the fulfillment of condition (1) may lead to considerable thermal perturbation of both real (Re ε) and imaginary (Im ε) parts of ε:

$$\text{Re } \varepsilon = \varepsilon_L (1 - \frac{\Omega_L^2}{\omega^2 + \nu^2(T_e)}) \tag{2}$$

Optoelectronics for Environmental Science, Edited by S. Martellucci and
A.N. Chester, Plenum Press, New York, 1990

$$\text{Im } \varepsilon = \frac{\varepsilon_L \, \Omega_L^2 \, \nu(T_e)}{\omega[\omega^2 + \nu^2(T_e)]}$$

here ε_L is the dielectric permittivity of the crystal lattice. These variations near the Langmuir resonance provide deep modulation of both amplitude R and phase φ of the complex reflection coefficient $R \exp(-i\varphi)$. Thus periodic heating of the carriers will modulate the continuous flow of incident radiation via oscillations in reflection and transmission. Herein, one may note different tendencies in such modulation depending upon the sign of derivative $(\partial\nu/\partial T_e)$, which is determined, in turn, by carriers scattering in the semiconductor plasma.

Unlike the traditional method of modulating the dielectric properties of semiconductors due with carrier generation by means of powerful laser irradiation, the proposed approach uses the modulation of reflection and transmission coefficients by thermal perturbation of the carrier collision frequency $\nu(Te)$, the density of carriers remaining constant. The characteristics time of reflection coefficient thermal tuning τ_R is determined by the carrier temperature relaxation time τ_T, the crystal lattice temperature being changed insignificantly. The relaxation time τ_T, e.g. for n-type III-V semiconductors such as InSb or GaAs, may be as small as $\tau_T \leq 10^{-9}$ sec.

Another feature of this tunable reflection is connected with the control of carrier heating by means of an external magnetic field. This field may be used to transition from the continuous regime of thermal variation of reflection coefficient to the bistable one, and vice versa. Since the power, used for thermal modulation is small, the ohmic heating of plasma carriers by means of an electric current proves to be suitable for such reflection tuning. In view of the simplicity of the related experimental set-up this case will be analyzed below. The principal element of such a set-up is a thin semiconductor film, which may be treated as a mirror with a tunable reflection coefficient. The potential for high speed amplitude-phase modulation and shaping of the reflected far IR radiation using such a mirror will be analyzed below.

II. THE PHYSICAL BASIS FOR THERMAL TUNING OF FAR IR REFLECTION NEAR A SEMICONDUCTOR LANGMUIR RESONANCE

Let us consider a simple model for the reflection of far IR waves from an n-type semiconductor with free carriers, the carrier energy relaxation being dominated by scattering from ionized impurities; the properties of this model, which are close to those of InSb at low temperature range $T_e \cong 10°K$, attract our attention due to the comparatively strong dependence[1] of carrier collision frequency on the carrier temperature T_e:

$$\nu_i = (T_e) = \nu_{io}/\pounds^{3/2}$$

$$\pounds = T_e/T_{eo} \tag{3}$$

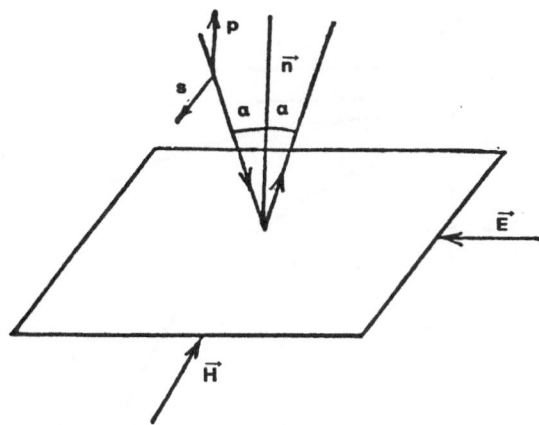

Fig. 1. The schematic set-up for tunable modulation of
reflected radiation: (1) the continuous
incident radiation beam, (2) the beam,
modulated due to reflection from semiconductor
film M; the heating electric field $\vec{E}$ is
directed parallel to the film's surface; $\vec{n}$ is
the normal to this surface. The vectors $\vec{s}$ and
$\vec{p}$ identify the s- and p-polarization of the
incident beam. The magnetic field $\vec{H}$ is
directed parallel to the film's surface and
orthogonal to the electric field $\vec{E}$.

Here $\pounds$ is the normalized value of carrier temperature, and T_{eo} and ν_{io}
are the unperturbed values of parameters T_e and ν_i. The temperature
dependence $\nu_p = \nu_p(T_e)$ in the case of polar scattering is more weak[2]:

$$\nu_p(T_e) = \nu_{po}/\pounds^{1/2} \tag{4}$$

The substitution of Eqs. (2) and (3) in Fresnel's formulae determines
the complex reflection coefficient for s- and p-polarized waves, α being
the angle of incidence (Figure 1)

$$R_s \exp(-i\phi_s) = \left[\cos\alpha - (\varepsilon - \sin^2\alpha)^{1/2}\right]/\left[\cos\alpha + (\varepsilon - \sin^2\alpha)^{1/2}\right]$$

$$R_p \exp(-i\phi_p) = \left[\varepsilon\cos\alpha - (\varepsilon - \sin^2\alpha)^{1/2}\right]/\left[\varepsilon\cos\alpha + (\varepsilon - \sin^2\alpha)^{1/2}\right]$$

$$\varepsilon = R_e\varepsilon + Im\,\varepsilon \tag{5}$$

this equation permits one to investigate the influence of carrier heating
on amplitude $R_{s,p}$ and phase $\phi_{s,p}$ modulation in the model given. The
fulfillment of the Langmuir resonant condition

$$\omega = \Omega_L = (4\pi e^2 N/\varepsilon_L m^*)^{1/2} \tag{6}$$

where N and m* are free carrier density and their effective mass, for far

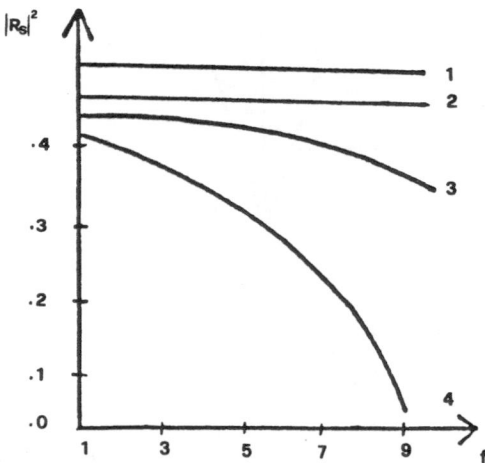

Fig. 2. The reflection coefficient R_s versus the carrier temperature £ in n-type InSb (α=45°, T_{eo}=78°K). The curves 1,2,3,4 relate to parameter A values 2.10^{-2}; 0,1; 0,5; 1, respectively.

IR and submillimeter radiation (ω=10^{12} - 2.10^{13} rad.s^{-1}) in the low temperature range $T_e \cong 10$°K may be provided by N=10^{14} - 5.10^{15} cm^{-3}. These values, which are greater than the values related to the equilibrium ionization of InSb at the temperature $T_e \cong 10$°K, may be reached using semiconductor doping with ionization impurities. The result of doping may be described by means of parameter A

$$A = N/N_2 \qquad (N_2 = m^* \varepsilon_L \, \omega^2/4\pi e^2) \qquad (7)$$

here the value of N_2 is determined from Eq. (6). The effectiveness of thermal tuning of such a doped semiconductor film reflectivity grows as the parameter A tends to unity (Figure 2). A similar resonance effect is manifested in the thermal steepening of spectral dependence on reflection coefficient $R^2(\lambda)$ near a plasma absorption edge (Figure 3). One may note the considerable (15-20 times) growth of the value of derivative ($\partial R^2/\partial \lambda$) in this resonant region.

The growth of temperature in this simplified model is limited, for several reasons. First, the concept of elastic collisions[3] is valid only so long as the carrier temperature remains below the Debye temperature T_D; e.g., T_D for InSb is equal to 240°K. Moreover, carrier scattering on ionized impurities proves to be the dominant mechanism of scattering only in the low temperature regime. Therefore, the normalized values of carrier temperature £ attained due to heating are restricted: £ $\cong$ 5 to 7. However, even such limited variations in temperature £, caused by small heat energy expenditure, may lead to effective tuning of the amplitude - phase structure of the waves reflected from the semiconductor surface. These variations depend essentially upon the angle of incidence α, for a constant

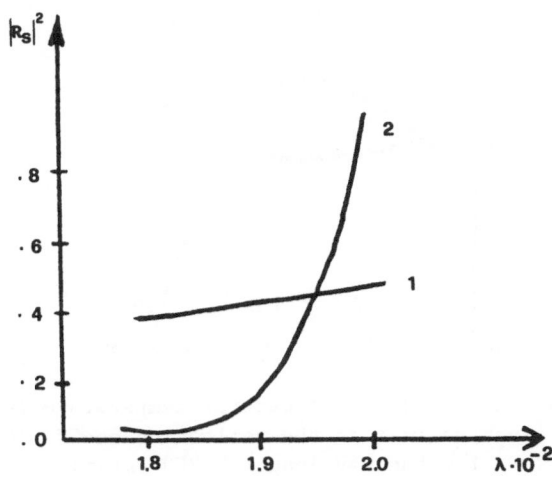

Fig. 3. The influence of carrier heating on reflection
coefficient R_s for n-type InSb ($\alpha=45°$) in the
resonant case (A=1); the curves 1 and 2
correspond to the values f=1 and f=7.

thermal perturbation f>1:

i)   the angular dependence of the reflected wave intensity $R^2 = R^2(\alpha)$ is
      strengthened due to carrier heating (Figure 4a). One may see from this
      graph that the depth of intensity modulation may be considerable
      ($R \to 0$) even for s-polarized waves ($R_s(\pi/4) \to 0$), although such tendency
      ($R \to 0$) in the idealized model of an absorptionless medium ($Re \varepsilon >0$,
      $Im \varepsilon \to 0$) is possible only for p- polarized waves near Brewster'angle[1]
      $\alpha_B = arc\ tg \varepsilon^{1/2}$;

ii)   the dependence of the phase shift ψ between the incident and re-
      flected wave upon the angle of incidence $\psi = \psi(\alpha)$ is also modulated due
      to carrier heating (Figure 4b); and,

iii) the thermal perturbations of amplitude R and phase   values for s-
      and p-polarized waves are different, Eqs. (5). This difference leads
      to a complicated deformation of polarization structure of the
      reflected wave, in both s- and p-polarized components.

      These examples illustrate the possibilities of Langmuir resonant
thermal tuning of amplitude, phase and polarization of the reflected wave.
The reflection coefficient, which is controlled by such heating, is
considered to be time-independent due to the stationary value of the
carrier temperature. In addition, the continuous flow of radiation
reflected from such a surface produces another continuous flow of radiation, although the amplitude and polarization are changed after the
reflection. However, in order to transform the continuous radiation flow
to a train of shaped pulses, it is necessary to produce temporal variations of the carrier temperature. This goal stimulates us to consider the

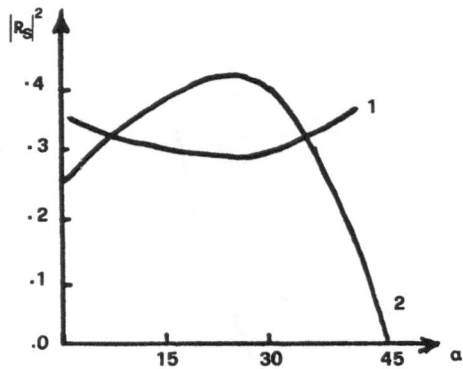

Fig. 4. The influence of carrier temperature in
n-type InSb on the reflection coefficient
R_s for the wavelength $\lambda=200\ \mu m(A=1)$.
R_s values are plotted versus the angle
of incidence α; curves 1 and 2 relate
to the cases $f=1$ and $f=7$, respectively.

problem of non-stationary modulation of semiconductor plasma reflective
properties.

III. MAGNETIC CONTROL OF STABLE AND BISTABLE REGIMES IN THE THERMAL AMPLITUDE-PHASE MODULATION OF A REFLECTED BEAM

Non-stationary processes in the reflected wave beam, described by
temporal thermal variations of the complex reflection coefficient
$R\exp(-i\varphi)$, depend upon the regime of carrier heating. The equation
describing the homogeneous ohmic heating of carriers by an external
alternating electric field $E=E_o \sin \Omega t$ (Figure 1), may be written in a
dimensionless form[1]

$$\frac{\partial f}{\partial \eta} + f - 1 = a^2\, F(f)\, \sin^2 \Omega t \tag{8}$$

Here η and a^2 are a normalized time variable and an energy parameter:

$$\eta_2 = t\, /\, \tau_T$$
$$a^2 = (E_o/E_p)^2$$
$$E_p^2 = 3m^*k\, T_{eo} (\Omega^2 + v_o^2)\, 2e^2; \tag{9}$$

the temperature relaxation time for carriers τ_T depends upon the total
unperturbed frequency of collisions v_o, all scattering mechanisms being
taken into account, and the sound velocity u in the semiconductor[2]

$$v_o = v_{po} + v_{io}$$

280

$$\tau_T = k\, T_{eo}/m^* \, u^2 \, \nu_o;$$ (10)

here e is the carrier charge, k is Boltzmann's constant, and E_p is the so-called "characteristic plasma field". The function $F(f)$, connected with plasma conductivity, depends upon the temperature £ via the collision frequency $\nu(£)$; for the field geometry shown in Figure 1 this function may be written in the form:

$$F(£) = \frac{1 + x^2}{2}\left[\frac{1}{(x-y)^2 + \nu^2(£)\nu_o^{-2}} + \frac{1}{(x+y)^2 + \nu^2(£)\nu_o^{-2}}\right]$$

$$x = \Omega\, \nu_o^{-1}$$

$$y = \omega_H\, \nu_o^{-1}$$ (11)

where ω_H is the carrier gyrofrequency associated with an external magnetic field H:

$$\omega_H = eH/m^*c$$ (12)

The temperature dependence $\nu(f)$ in Eq. (11), with both carrier scattering mechanism Eqs. (3) and (4) being taken into account, may be written as

$$\nu(£) = \nu_o\, £^{-3/2}(1 - p + p£)$$ (13)

$$p = \nu_{po}\, \nu_o^{-1}$$

Eq. (8) must be solved with an initial condition:

(14)

$$£\big|_{\eta=0} = 1$$

Crystal lattice heating is neglected in the energy balance Eq. (8) due to the considerable difference in heat capacity values between the carrier gas and the lattice. This assumption is justified by detailed calculation[4] of lattice heating in such processes.

Some features of such heating may be outlined in the framework of the stationary model ($\partial f/\partial\eta = 0$), related to the case of constant heating field E_0=const. The temperature f, in the absence of a magnetic field (H=0), may be calculated from the energy balance Eq. (8)

$$(£ - 1)\left[x^2 + (1 - p + p£)^2/£^3\right] = a^2\,(1 + x^2).$$ (15)

Two types of collision, Eq. (13), are taken into account in Eq. (15).

An important property of Eq. (15) is the possibility of a bistable solution $f=f(a^2)$; the conditions for bistability depend upon the parameters x (Eq. 11) and p (Eq. 13). In the idealized model p=0, well known in gaseous plasma physics[1], such bistability arises in the low frequency range ($\Omega<0.25\nu_o$, x ≤ 0.25). The appearance of two temperatures states in this case is due to the existence of two different tendencies in the

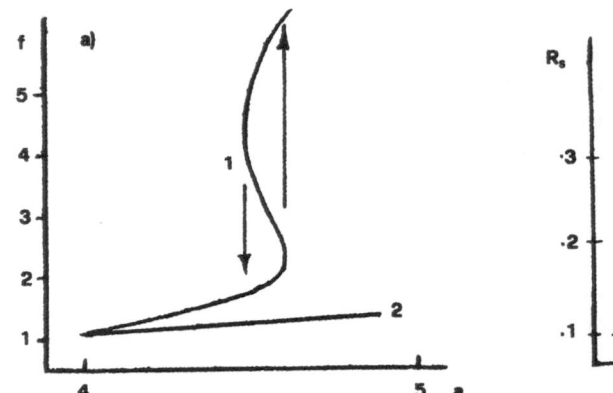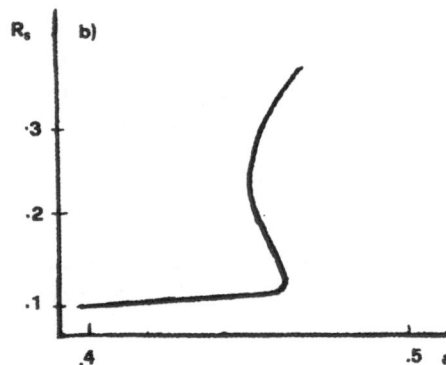

Fig. 5. Magnetic tuning of thermal bistability onset in reflection of s-polarized wave (A=1, T_{eo}=78 K) from n-type
InSb. (a): the carrier temperature f is plotted against
the energy parameter a (Eq. 9); curves 1 and 2 illustrate
the hysteretic heating regime (p=0.3, y=0.14 < y_{cr}=0.16)
and the monotonic one, formed due to magnetic field
growth up to parameter y value y=1 > y_{cr}=0.16.

dependence of plasma conductivity ∂ upon collision frequency:

$$\partial = \frac{e^2 N}{m^*} \frac{\nu(f)}{\Omega^2 + \nu^2 (f)} \tag{16}$$

initially, the decrease in frequency $\nu(f)$ due to heating in the case
$\nu_0 \geq 4\Omega_0$ leads to an increase in ∂ : $\partial \cong \nu^{-1} \cong f^{3/2}$; however, the continuing
decrease of $\nu(f)$ leads to an opposite inequality: $(\nu(f)<\Omega)$; in this case
the conductivity decreases: $\partial \cong \nu \cong f^{-3/2}$.

From a formal viewpoint, the foregoing reasons for bistability regime
formation might appear in the case under consideration due to the
considerable difference between the typical values of $\nu \leq 10^{12}$ sec^{-1} and
$\Omega \ll$ 10 GHz (x<<1). However, such a difference leads to extremely high
values of the temperature growth ($f \geq 30$) necessary for the bistability
appearance; these values are beyond the scope of the model presented here.
On the other hand, the range of continuous temperature growth from the
initial point $f = 1$ (Eq. 14) up to the value $f=f_{cr}$, calculated from the
f - derivative of a^2, is restricted by the value f_{cr}=1.5 in the model p=0.
Thus, this method of ohmic heating proves to be insufficient to obtain the
above-mentioned values $f \leq 7$ in the problem under consideration.

However, the problem of continuous heating up to values $f \leq 7$, as in
Figures 2 - 4, may be solved simply by using a magnetic field, provided
that the carrier gyrofrequency is high enough. In the model p=0 the
bistability phenomena vanish in the case

$$\omega_H > 0.25 \nu_o \tag{17}$$

282

A decrease in gyrofrequency ($\omega_H \leq 0.25\ \nu_o$, $y \leq 0.25$) leads to the appearance of the bistability regime. The same tendencies are manifested in the more realistic model as well, when both types of scattering mechanisms (Eq. 13) are taken into account. However, the boundary value $y_k = 0.25$ (Eq. 17), related to the case $p=0$, decreases due to growth in the parameter p. The analysis of Eq. (15) shows that the bistability regime vanishes when $p \to p_{cr} = 2.3^{-1}$. An example of such bistability in the carrier temperature is illustrated in Figure 5a; the related hysteretic jumps of amplitude and phase of the complex reflection coefficient are shown in Figures 5b and 5c.

Thus, a magnetic field may be used for switching the reflection regime from hysteretic to a non-hysteretic one, and vice versa, by means of use of Eq. (17) or an opposite one. Herein the light carriers in n-type InsB ($m^* = 0.013\ m_0$) permit one to operate such switching processes by a relatively weak magnetic field $H \leq 0.5$ kGs. One may emphasize the following features of such magnetic control of thermal tuning processes:

i)   the very narrow range of energy parameter a^2 values defining the bistability regime;
ii)   the sensitivity of the bistability regime to small variation of magnetic field;
iii) the energy gain in the process of reaching some temperature in the hysteretic regime, in comparison with non-hysteretic one (Figure 5a).

These hysteretic jumps may be used for switching the intensity of a continuous reflected beam from one branch of the S-like curve (Figure 5b) to the other one.

On the other hand, the shaping of reflected pulses by means of periodic heating may be analyzed by means of the non-stationary Eq. (8). Let us emphasize that the temperature relaxation time for carriers τ_T, Eq. (10), is only 10^{-9} to 10^{-10} sec. The period of the alternating heating wave $T_0 = 2\pi\Omega^{-1}$ is usually restricted: $T_0 \leq 10^{-9}$ sec. Thus, the dynamics of the temperature relaxation process in the heating electric field $E = E_0 \sin\Omega t$ depends upon the ratio τ_T/T_0.

IV. HIGH SPEED SHAPING OF REFLECTED WAVE BEAMS

Unlike traditional magnetooptical effects, arising from the direct influence of magnetic field on refractive index and manifested even in an absorptionless medium, this chapter concerns an indirect influence of magnetic field on reflection coefficient, produced due to this field's influence on the dissipation of the heating current. Here the anisotropy of semiconductor reflection properties, caused by this magnetic field, proves to be insignificant for the wave under consideration, provided that certain condition are fulfilled, concerning the smallness of the ratio ω_H/ω or the reflection geometry. A simple example of this kind is given by an s-polarized wave, when the electric vector of this wave $\vec{E}_\sim$ is parallel to an external magnetic field $\vec{H}$ (Figure 1). The magnetic field in this geometry has no direct influence on the reflected wave, for arbitrary ratio ω_H/ω. However, such a magnetic field may lead to shaping and deflec-

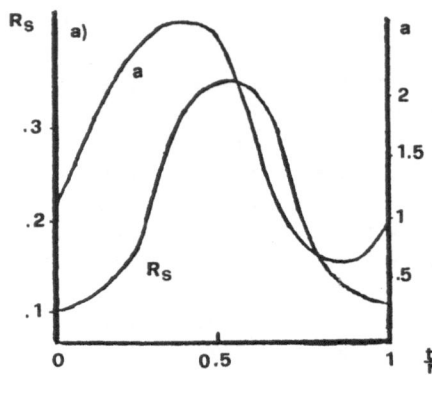 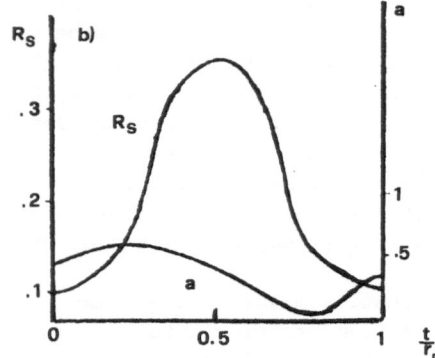

Fig. 6. The shaping of symmetrical reflected pulses with the envelope
$R_s(\eta)$, the incident s-polarized wave amplitude being equal to
unity, by means of monotonic (Fig. a: y=1 > y_{cr}=0.16) and
hysteretic (Fig. b, y=0.14 < y_{cr}=0.16) heating regimes,<,
v=0.7; ν_o=3Ω_L ; α=45°; τ_R=30τ_T.

tion of reflected beams. Thus, there are interesting possibilities for
the controlled formation of symmetrical shapes of reflected pulses, with
special choice of the heating pulse envelope.

These phenomena may be illustrated by means of non-stationary
computer simulation of Eq. (8). Let the temperature values, determined
from the stationary equation ($\partial f/\partial \eta$ = 0), be equal to f_{st}. Then the
amplitude of temperature oscillation described by Eq. (8) achieves values
close to f_{st}, when the heating wave period T_0 is much greater than the
temperature relaxation time τ_T: $T_0 \cong (20-30)\tau_T$. The computer simulation
permits us to choose the envelope of the heating field a(η), responsible
for forming the given envelope $f(\eta)$ which stipulates, in turn, the
reflected pulse shaping. This pulse is characterized by the functions $R(\eta)$
and $\varphi(\eta)$, the amplitude of the incident wave being equal to unity. Thus we
may discuss a reflected pulse shape with smoothened edges, described by two
constants, A and B:

$$R(\eta) = A - B \cos (2\pi\theta\eta) \qquad (\theta = \tau_T/T_0) \qquad (18)$$

These symmetrical pulses $R(\eta)=R(-\eta)$ may be produced by both continuous
(Figure 6a) and hysteretic (Figure 6b) heating regimes. It is interesting
to note that the symmetrical shaping of the reflected pulse in the range
of bistable dependence $f=f(a^2)$ arises from an asymmetrical envelope a(η).
The calculations show that a harmonic envelope a(η)$\cong$sin($2\pi\theta\eta$) may cause[5]
the formation of a complicated asymmetrical profile $f(\eta)$.

V. CONCLUSIONS

The remarkable properties of such a tunable mirror for far IR
radiation arise from the use of very thin semiconductor films. The films
thickness d may me determined as an absorption length for the incident
radiation

284

$$d = c/\omega x; \tag{19}$$

here the imaginary part of the complex refractive index n+i may be calculated via the complex dielectric permittivity ε (Eq. 2):

$$\varepsilon = (n + ix)^2 \tag{20}$$

The d value in the range under consideration proves to be very small: $d \cong (15-30)$ nm; this parameter restricts the semiconductor film thickness. The power Q for free carrier heating in the unit volume

$$Q = \frac{3}{2} N k T_{eo} (\pounds - 1) \tag{21}$$

does not exceed $Q \cong 10^{-5}$ W cm^{-3}. The volume of semiconductor film, due to the aforesaid small thickness, is very limited: $V \leq 0.1$ cm^3; thus, the total power used for such element heating is as small as 10^{-5} W.

The potential for soft tuning of duration, shape and polarization of the reflected signals, high speed of action and low energy expenditure for the 5 to 7 times growth in carrier temperature growth, necessary for these effects, permits us to consider these phenomena for tunable and adaptive optics in the far IR and submillimeter ranges.

VI. ACKNOWLEDGEMENTS

The author thanks Prof. F. Kneubuhl and Prof. S. Martellucci for their interest to these problems.

REFERENCES

1. V. L. Ginzburg and A. A. Rukchadze, "Electromagnetic Waves in a Plasma", Moscow, Nauka (1975) (in Russian)
2. K. Seeger, "Semiconductor Physics", Springer Verlag, Wien, New York (1973)
3. N. V. Karlov, E. A. Sissakyan and A. B. Shvartsburg, Phys. of condens. matter, B 35, 107 (1985)
4. N. A. Antoniuk, A. V. Shepelev and V. A. Skorik, Dokl. Akad. Nauk., 307, 21 (1989) (in Russian)
5. A. B. Shvartsburg and M. A. Zuev, Computer Opt., 4, 53 (1989) (in Russian)